TOXIC LAKE

Toxic Lake

*Environmental Destruction and the
Epic Fight to Save Onondaga Lake*

Thomas Shevory

NEW YORK UNIVERSITY PRESS
New York

NEW YORK UNIVERSITY PRESS
New York
www.nyupress.org

Library of Congress Cataloging-in-Publication Data
Names: Shevory, Thomas C., author.
Title: Toxic lake : environmental destruction and the epic fight to save Onondaga Lake /
Thomas Shevory.
Description: New York : New York University Press, [2023] | Includes bibliographical
references and index. | Summary: "Onondaga Lake is sacred territory for members of the
Haudenosaunee Confederacy. But by the mid-twentieth century, it was dubbed "the most
polluted lake in America." The most expensive cleanup effort in American history was
initiated in the 1990s, which, in turn, generated a new set of controversies"—Provided by
publisher.
Identifiers: LCCN 2022059285 | ISBN 9781479815678 (hardback ; acid-free paper) |
ISBN 9781479815685 (paperback) | ISBN 9781479815708 (ebook) |
ISBN 9781479815715 (ebook other)
Subjects: LCSH: Water—Pollution—New York (State)—Onondaga Lake. | Sewage disposal
in rivers, lakes, etc.—New York (State)—Onondaga Lake. | Water quality management—
New York (State)—Onondaga Lake. | Environmental protection—New York (State)—
Onondaga Lake.
Classification: LCC TD424.35.N7 S54 2023 | DDC363.739/40974765—dc23/eng/20230111
LC record available at https://lccn.loc.gov/2022059285

This book is printed on acid-free paper, and its binding materials are chosen for strength
and durability. We strive to use environmentally responsible suppliers and materials to the
greatest extent possible in publishing our books.

Manufactured in the United States of America

10 9 8 7 6 5 4 3 2 1

Also available as an ebook

CONTENTS

Introduction

A Toxic Tour

Onondaga Lake, a 4.6 square-mile lake in the Syracuse area, was once deemed "the most polluted lake in America."[1] It's now in better condition thanks to several decades' worth of political organizing, scientific investigation, and a half-billion-dollar restoration effort. The lake is not fully restored, however, and what that term even means is the subject of intense debate. But whatever the specifics of its current status, there's little doubt that Onondaga is no longer among the country's most polluted, and given possible contenders like the Berkeley Pit in Butte, Montana, it's not entirely clear that it ever was. I wanted to see for myself, so one day, traveling with a bicycle stuffed into the back of my 1993 Volvo 240 Station Wagon, I set out on a tour of Onondaga's contamination sites. My goal here was to gain a material sense of the breadth of Allied Chemical's contamination of Onondaga County, and to communicate that with the readers of this study. While the lake itself, and its immediate watershed, are at the center of this story, the reach of Allied's impacts stretches far beyond them.[2]

I first wanted to inspect the Tully Creek mudboils. To understand the nature of the boils, you need to know something about the Solvay process, a major contributor to Onondaga Lake's contamination. More detail about the process is provided later on, but, for now, it is important to know that it involved the production of soda ash, a component for multiple manufacturing needs since the nineteenth and twentieth centuries, glassmaking being primary among them. The production of soda ash requires several constituents, including large amounts of salt. Central New York has long been a hub of salt mining. The City of Syracuse is nicknamed "salt city," and there are still vast underground salt mines in the area, reaching underneath the bottoms of two of the Finger Lakes, Seneca and Cayuga.

Syracuse's more accessible salt deposits were mostly depleted by the middle of the nineteenth century, so when Solvay engineers went looking for salt, they had to turn southward. The area near small Tully Lake had some excellent underground deposits, but it was ten miles from the Solvay production facility. In order to transport the salt as efficiently as possible, the operators pumped water from Tully Lake into the salt deposits to create the brine that would be used in the Solvay process. The water pressure was such that pumps were unnecessary to move it up to the production facility. (Brining was a common—and now discredited—method of obtaining salt from underground deposits in the nineteenth century.)

The salt deposits lay underneath Onondaga Creek, which flows northward through the City of Syracuse and feeds Onondaga Lake. The creek cuts through the Tully Valley, an area described as having a "spectacular glacial topography."[3] At the southern end is Tully Lake, which sits on the Valley Heads Moraine, part of the St. Laurence Seaways Divide. As the salt was removed, fissures and cracks appeared in the hillsides surrounding the valley, and valley walls started to sink, allowing water to flow ever more rapidly into the caverns. While the plant was in operation, brine was removed from the mines so as to relieve the pressure generated by the operation. Once soda ash production stopped, pressure increased and salty, muddy water was forced upward into Onondaga Creek in a boiling swirl. Records of mudboils appear as early as 1899, but they increased dramatically in the 1990s. Ground subsidence (the sinking of landforms) also intensified. In 1993, part of the hillside on Bare Mountain (the western valley wall) gave way, destroying three homes and covering Tully Farms Road with twelve feet of mud.[4]

I had a map from the U.S. Geological Survey that showed where the boils were located, but not being familiar with the area, it took me a while to get my bearings. There appeared to be a road that went down to the creek almost exactly intersecting with the boils. I drove along through farmland to get there. This region of the state used to be the center of New York's dairy industry, but it has fallen on hard times, so many of the dairy farms were converted to commodity crops, such as corn. Still, it's a very beautiful area. The narrow road down to the creek came to a dead end just below a farmhouse. A bridge that once crossed over was destroyed by subsidence and never replaced. I got out

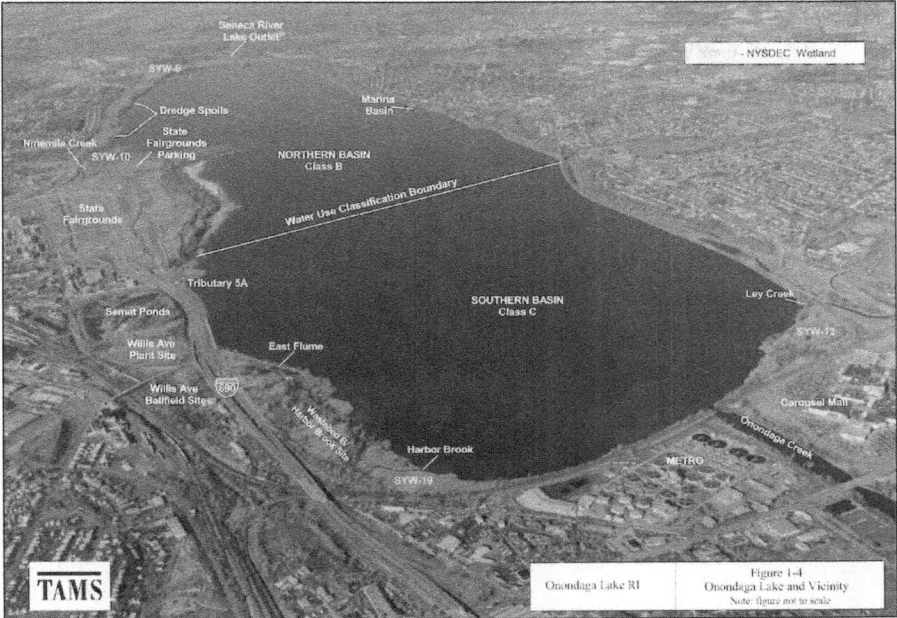

Figure 1.1. Onondaga Lake. Courtesy of the Honeywell Onondaga Lake Cleanup website.

of the car to confront an impenetrable thicket of brush. I was in a t-shirt and shorts on a hot summer day, and I didn't see any way down to the creek that didn't involve a machete, some protective clothing, and a fair amount of insect repellent. There was definitely no path. I figured I'd try the other side.

In the car, traveling back south, I spotted a public access point for fishing in Onondaga Creek. I checked it out, only to find much the same thing—an overgrown thicket that looked as though no one had passed through it in years. It had been a hot, dry summer, so my guess was that the creek was pretty dry and there wasn't much fishing happening. Perhaps in the spring, when the trout ran, there might be more activity and a better path.

I drove south and then turned back up northward on the other side of the creek. I knew from the map that I was traveling over the empty salt caves created by the Solvay process. On the way back north, I saw a sign for Solvay Road, named during the heyday of salt mining and soda ash production. Photos from the Solvay archives show strange, medieval-looking brine wells that pumped the liquid up, before it was funneled into the pipe that carried it to Solvay. The old photos capture mysterious and slightly haunting, almost sinister-looking structures. They look alien to such a verdant setting, but, when standing, provided evidence of the solution mining process taking place underneath the ground.

Solvay Road is only a mile or two long. At the end, I turned left, and went up northward along the other side of the creek. I spotted a narrow farm road and took it to a dead end once more at a thicket underneath which, somewhere, was Onondaga Creek as well as the source of the mudboils. I couldn't make my way down to it, but I also knew that a highway crossed the creek just to the north, so I decided to take a look from there.

I continued on my way and passed into the Onondaga Nation's land. The creek flows through here, but the silt caused by boils, land subsidence, and erosion has eliminated the trout that used to be plentiful. The Onondaga, or Hill Place people, who will be discussed later in greater detail, are the Indigenous people of what is now Central New York. They are members of the Haudenosaunee Confederacy, which dominated New York, Western Massachusetts, and points beyond for hundreds of years before the arrival of Europeans and the confiscation of their lands that occurred mostly in the period following the American Revolutionary War.

The Onondaga, being centrally located in the Haudenosaunee territory held an especially important place in the confederation. They were designated as "Keepers of the Fire," and their territory was used as a meeting place. A thousand years ago, Deganawida, or Peacemaker, a visionary from the northern shore of Lake Ontario, approached the Onondaga, and, with the aid of Hiawatha, established the confederacy after a long period of intense conflict. The Tree of Peace grew in the Onondaga territory, and the Great Peace was memorialized by Hiawatha via wampum beads that were, according to one account,[5] found on the bottom of Lake Tully. Hiawatha is said to have met Peacemaker, the originator of the idea for the Great Peace, on the shores of Onondaga Lake. The waters and grounds of this area are sacred to the Onondaga people, and those of the entire Haudenosaunee Confederacy. The current reservation is 9.6 square miles, or roughly 5% of the original Onondaga territory, the rest having been expropriated by the State of New York in the late eighteenth and early nineteenth centuries.

Figure 1.2. The Tully Brine Wells (n.d.). Reprinted with permission from the Solvay Public Library and Village of Solvay.

I found my way to where the highway intersected with the creek. There's a small parking space on the side of the road. I got out of the car and spent some time watching the very muddy creek flow under the bridge. When I eventually was able to interview Clan Mother Wendy Gonyea, she told me that when she was growing up, families would have picnics along the creek's shoreline. People would swim and fish the creek. Her brother still fishes from it.[6] But most of the trout have disappeared, and I, myself, would be reluctant to eat any fish caught here.

Next, I continued traveling northward into the city of Syracuse, New York. Founded in 1820, Syracuse sits in the center of New York State. The city proper has nearly 150,000 inhabitants; the metropolitan area over one-half million. Early on, its economic mainstay was the salt industry, but, like other upstate cities, it evolved into an industrial hub in the early twentieth century, and, as such, has faced economic challenges in recent decades. Syracuse University, established in 1870, is still a prominent institution of higher education.

On my way to the Village of Solvay, I had a couple of stops to make. First, I wanted to take a look at the Midland Avenue Regional Treatment Facility. The facility was designed to capture overflow from Syracuse's combined rainwater and sewage system, which can be overwhelmed during heavy rain and snowmelt events, causing backup into basements, and then overloading the main sewage treatment plant, resulting in the flow of raw sewage into Onondaga Lake. Onondaga County was under federal court order to deal with the overflow, and, in response, proposed building several interceptor facilities, such as the one on Midland Avenue. The facility was constructed in a predominantly African American community on Syracuse's south side. A grassroots movement to prevent an industrial facility from being sited in a residential community, and to prevent houses being demolished to accommodate it, gained a good deal of support from across the city, but was unsuccessful, and, in 2006, the facility was constructed. The county then proposed the Save the Rain Program, to use environmentally friendly means to reduce overflow problems and to eliminate the need for additional interceptor facilities, a solution that was approved by Federal District Judge Frederick Scullin in 2009.

The Midland Plant sits on Onondaga Creek and is a drab brick building with no windows. It's obvious where the houses were torn down when it was constructed. A bus garage of some kind, which provides another large industrial footprint in a residential neighborhood, sits next to it. My guess is that neither of these would have been sited here if this was a more affluent community, and one, it needs to be said, with fewer people of color.

Next, I drove up to the Rosamond Gifford Zoo, a forty-three acre complex at the top of a hill on the West Side of Syracuse. While more than 100 years old, the buildings and grounds were renovated in 1986, giving it a modern feel, and like many zoos, it has recalibrated itself as a conservation center, rather than simply a place to view confined, exotic animals. But I wasn't there to visit the aviary or to see the golden lion tamarins, the Chilean flamingos, or the Cuban whistling ducks. Rather, I was interested in several Save the Rain projects that were integrated into the zoo's acreage. To start, the parking lot is made of porous pavement, which doesn't look very different than conventional pavement, but it, along with green borders and dividers, retain significant amounts of runoff. Inside the zoo, there are swales (or vegetated channels) near the primate exhibit, the courtyard of which is also made of porous cement. The elephants, which are given several acres to roam, also have access to a green-roofed exhibit building. In addition, there's a stormwater wetland near the penguin exhibit.

Save the Rain projects such as these have been implemented within the central city, but they are dispersed and not always easy to find. My guess is that, unless you were actively involved with environmental or public health issues in the City of Syracuse, you probably wouldn't notice them. They are seamlessly integrated into the built environment, often adding aesthetic value, but not in an intrusive way. Unfortunately, this might result in underappreciation of the program, which now has national prominence and is a model for cities facing similar runoff problems. In Syracuse's case, the projects allow the county to halt the construction of more interceptors, saving money in the process.

Next, I headed into the Village of Solvay. With a population of just under 7,000 residents, it sits on the northwestern edge of its larger neighbor. I took Willis Avenue, the main street in town, southeast away from the town center to the lake. I passed the hulking Crucible Steel building on my left, now empty, and the strikingly blue lake on the right. A New York State Fairgrounds parking lot is near the trailhead where you can gain access to Restoration Way, a bicycle/walking path that now circles much of the lake. The parking lot is on a hill, the only hill on the lakeshore. The hill was created over decades, the result of Allied Chemical calcium chloride waste—residue left over from the Solvay process, which required carbon to generate carbon dioxide. The necessary carbon was drawn from massive limestone (calcium carbonate) deposits that are also part of the geological substructure of Onondaga County. Once the carbon was extracted, large amounts of calcium chloride, contaminated with a variety of heavy metals and organic chemical toxins remained. Calcium chloride wastebeds ring the southeastern end of Onondaga Lake, constituting the 312-acre hill I was on, and have been designated numbers 1 through 8 for purposes of remediation. These disposal sites operated from 1926 to 1944, but were closed when a dike failed, some of the material gave way, and a toxic mess flooded the streets of the Village of Solvay.

In the parking lot, I pulled my bike out of the back of the car. There were ten or fifteen other cars in the lot, as people were getting their bikes ready or walking over to the trail. The trail wasn't crowded—people were scattered enough that I didn't feel unsafe riding along it in the midst of a pandemic. It's a wide and well-maintained trail that travels along the top of the calcium chloride pile that forms a peninsula, and then close to the lakeshore. A hundred years ago, it would have been impossible to ride a bicycle or walk here. Before Solvay process wastes were dumped, the whole area was part of the lake. The entire peninsula is an artificial creation and a testament to twentieth-century waste disposal practices. A half-mile or so along the trail, I encountered the St. Joseph's Health Amphitheatre. The venue was one of the early development projects initiated as the lake restoration efforts went forward. It has 17,500 seats and is now a popular concert venue.

After the amphitheater, the trail curves to the right and heads downward, and then back to the left and over Nine Mile Creek, the second largest creek flowing into the lake. Nine Mile was highly contaminated during the Allied Chemical years. When wastebeds 1 through 8 were closed, beds 9 and 11 were opened up, just northeast of the creek. To the southeast of these were the noxious wastebeds 12–15, where brine purification sediments were dumped, along with mercury-contaminated wastewater, boiler water purification wastes, and ash. From 1981 to 1986, wastewater from the Metro Sewage Plant was sent here, to precipitate out phosphorous, which would bond with calcium chloride to keep it from entering the lake where it could create ecological havoc. The water was then funneled into Six Mile Creek. Also, for a period of time in the 1980s, sewage sludge from the Metro plant was deposited here. The south end of these wastebeds touch the West Flume, a small creek that was dredged to use as a receptacle for various effluents. The flume flows into Geddes Brook, which in turn flows into Nine Mile Creek, creating an intertwined tributary system of aqueous contamination. Lake restoration efforts have included some dredging at the mouth of Nine Mile Creek, where decades worth of contamination had collected.

As I continued northward along the path, the village of Lakeland now to the east, the trees became denser, and it felt as though I were in actual forest, with the lake only a few feet to the right of the path. The land here was largely uncontaminated by Solvay wastes, although it didn't entirely escape sewage contamination. It almost has a wilderness feel, which is unfortunately disrupted by the noisy traffic traveling along I-690, which runs a couple of hundred feet to the east.

At the northern end of the lake, there's another parking space for visitors and a narrow bike path over the Long Branch Road bridge. On the other side, you enter into Long Branch Park, which was developed as an amusement park in 1882, part of the whole system of resorts that populated the shores of the lake. All were eventually closed as trolley and steamboat service ended and the lake became increasingly smelly and unattractive. Long Branch Park closed as an amusement park in 1925, but it has remained open as a public park with picnic areas, pavilions, and a sledding hill.[7] According to Clan Mother Wendy Gonyea, this is the area where the Onondaga believe that Peacemaker first brought his message of peace to the warring factions of the Haudenosaunee.[8]

The Park has a walking path and a narrow road for cyclists. The road follows the lake about 100 feet to the west, eventually passing by houses in the Village of Liverpool. The land on this side of the lake was never contaminated by Solvay, although you can see the industrial skyline and houses of the village of Solvay across the lake a couple of miles away, a reminder of why, even on a beautiful day like this, no one can be seen swimming in Onondaga's waters (despite the fact that they have, for the most part, been deemed safe by regulatory authorities).

At Onondaga Lake Park, about halfway down the western side of the lake, the path ends, and I rode out onto the Onondaga Lake Parkway. Onondaga Lake is virtually surrounded by interstate highways, and at the very southeastern corner of the lake, there's a tangle of entry and exit ramps as the road intersects with another interstate, I-81, which runs north to the Canadian border and south into Pennsylvania. I made my way carefully, trying not to get run over by fast moving traffic, over to an entrance to Destiny Mall, passing over Ley Creek, another Onondaga tributary. The area around Ley Creek was contaminated by a General Motors plant that used to operate nearby. It closed in 1993, but in 2019 PCBs were discovered in the backyards of nineteen homes close to the site, which are now undergoing remediation efforts.

The mall that I approached had opened in 1990 as the Carousel Center. Over time the ownership changed, expanding and adding new tenants. It was redubbed Destiny USA in 2012. The mall was built on a toxic site known as Oil City, which, as the name implies, was an oil tank yard. Remediation of the site involved collecting the toxic petroleum residues and other chemicals, wrapping them in sheets of plastic and burying them at the construction site, where they still sit. During the period of mall expansion, when new pilings were driven into the area adjacent to the original mall, contractors were warned to be careful and not puncture the plastic sheeting, because it would release tons of toxic contaminants.

A driveway at the north end of the mall (and the southern tip of the lake) ends at Onondaga Creek, the other main tributary into the lake, and the same creek that features the mudboils that I had unsuccessfully tried to locate earlier. The creek runs through the city of Syracuse and was used for direct sewage disposal in the nineteenth century, until the first sewage treatment facility was constructed in the early twentieth century. To get across the creek I had to make my way on a bike path up to Hiawatha Boulevard.

Once over the creek, I crossed to the north side of the boulevard in front of the Metropolitan Sewage Treatment Plant. I document the history of this plant later, but for the time being it is worth noting that, while it is now a state-of-the-art tertiary treatment facility, its predecessors were inadequate to meet the needs of a growing twentieth-century city. One problem was the city's decision to combine storm water and sewage into one system. This was a practice followed in many eastern and midwestern cities because it allowed for sewage to be diluted via rainwater and snow runoff, decreasing its density and smell. The result, however, was that during heavy rain events, the sewage plant would be overwhelmed, and raw sewage would be diverted directly into Onondaga Lake.

When planning its first sewage treatment plant, the city had a choice between an Imhoff tank and a Dorr clarifier. The Imhoff system involves bacterial digestion of the sewage, facilitated by aeration, and a settling system to collect solids once the digestion has been completed. The Dorr system is just a settling tank, with no aeration. The city chose the cheaper Dorr system and saved 1.5 million dollars, 21 million in today's terms. It has paid the price for this again and again, however, in lake contamination and numerous expensive attempts at upgrades. Because the Dorr system worked so poorly, the Solvay wastebeds were used as a kind of filter; sewage was diverted to them before running it into the lake. By the 1940s, Onondaga Lake had become a virtual cesspool, and an odd and destructive symbiosis was established between Allied Chemical and Onondaga County, as officials contended that Solvay wastes essentially cleaned the sewage, or at least made it smell less bad. As a result, officials wanted Solvay to keep piling up its wastes, so they could be used to disguise the extent of sewage contamination from the city. Only after the plant closed was the county forced to deal with the sewage issue more directly. Still, it took two more decades, after Allied's shuttering, before the current plant was brought online.

After the Metro plant, there is a vacant fenced-off area at the southwestern end of the lake. This was the site of Roth Steel, a metal shredding facility that had been in business for 100 years. In 2008, the company settled with the State of New York for improper disposal practices. A steel shredding operation generates tremendous amounts of what is known in the business as "shredder fluff." This is the extraneous non-metallic, sometimes toxic, material that can't be recycled. Roth Steel apparently dug a hole and dumped large amounts of this material into it and did so for years. An employee eventually outed the company; the state became involved; and the company was fined $150,000 dollars and required to remediate the site. In 2015, Roth declared bankruptcy while still owing Onondaga County $60,000 in economic development money.

Hiawatha Boulevard ends at Erie Boulevard. I turned right onto the latter. Erie Boulevard, in turn, ends at Willis Avenue, which runs west and then north into the Village of Solvay and finally east back toward the lake. I document some of the social and economic history of Solvay in chapters 2 and 3. Because my goal was to circumnavigate the lake, and since I'd spent a fair amount of time in the village previously, I turned right onto Willis Avenue heading east.

On the right were some dilapidated houses. To the left is a vacant area, fenced off with chain link and barbed wire. On the other side of the fence is a mound, approximately fifty feet high. On top is an old reservoir that once held the brine that was piped up from Tully, where I had stopped earlier in the day. I looked around for a gap in the fencing so that I could climb up to get a peek, but with no luck. I stopped to get my bearings and then trundled over a bridge, underneath which are railroad tracks.

In the heyday of industrial activity, these railroad yards buzzed with activity. That is not the case today. The site of the former Solvay/Allied main plant was on the left after I crossed the bridge. This facility generated most of the waste material that is piled up in various sites around the lake and surrounding areas. The former site is now occupied by The

Figure 1.3. Allied Brine Reservoir (1986). Courtesy of the Library of Congress photo archive.

West Rock Company, makers of cardboard packing materials. Wastewater from that plant is treated with bacteria to remove the carbonaceous material that's left in it. The methane off-gases are then sent back to the production facility, and some of the water is recycled into boilers. The rest is sent over to the Metro treatment facility. The West Rock facility marks significant environmental progress, given how differently its wastewater is processed from past practices, which mostly involved dumping waste materials into the surrounding land and water.

Over the bridge I could see the lake and a bird's-eye view to the southeast of the area where I parked my car. Wastebed B, on the southwest corner of the lake, was the main repository for Solvay wastes from 1908 to 1926. From 1959 to 1965, Metro sewage sludge was sent there to minimize its smell. To the west of its northern tip, the East Flume was excavated as a means to funnel waste deposits into the lake itself, which was one of the primary mechanisms by which mercury drained into it. Abutting Wastebed C is Wastebed B, a.k.a., the Willis Avenue Ballfield site. Here, Allen-Moore diaphragm cells, used in the production of chlorine and caustic soda, and heavily contaminated with mercury, were deposited along with a slew of other chemical wastes and debris. In 1960, Allied Chemical covered it with dirt and made it into a ball field.

The Willis Avenue plant, which produced the caustic soda that created the mercury pollution, lies just north of the ball field. Chlorine was a byproduct of caustic soda production, so the company put it to use by combining it with benzene, to create a chemical that can be used in the production of herbicides, dyes, paints, paint removers, and pharmaceuticals. As a result, the area around the Willis Avenue plant became heavily contaminated with chlorinated benzenes.

The Semet Residue Ponds, which are just north of the Willis Avenue plant, were established to collect wastewater from the production of benzene, toluene, and xylene. Before remediation of the site began in 2008, the ponds contained twenty million gallons of concentrated liquid toxins. Between the ponds and the Willis Avenue plant, Allied Chemical dredged Tributary 5A, which became another route for wastes from the plant to enter into the lake and settle on the lake bottom.

Because of the presence of mercury, chlorinated benzenes, and other contaminants flushed into the lake from these facilities, Honeywell, which inherited the site when it was absorbed by Allied, was required to dredge and cap parts of it as part of its cleanup plan. The company also constructed a facility to capture and filter groundwater, to prevent further contaminants from draining into the lake. They constructed a barrier wall that reached along the shoreline, several feet into the lake itself, and, in doing so, reduced the lake's size. Attorneys for the Onondaga Nation objected to the plan, claiming accurately that it did not restore the lake to its original condition.

After the bridge and at the end of Willis Avenue, I turned left onto State Fair Boulevard, which runs parallel to I-695 and the shoreline of Onondaga Lake. On the left is the Crucible Steel building that I passed earlier in my car. The roots of the company go back to an iron works that was established in 1870, but the works became Crucible after a merger in 1900. Steel production stopped in Syracuse in 2009, under the shadow of a Chapter 11 bankruptcy. Wastes from the Crucible plant, like the Solvay facilities, went into Onondaga Lake. Chromium, grinding dust, slag, boiler sludge, and other kinds of debris were deposited to create a small peninsula that juts out into the lake, just below the main Solvay wastebeds.

In my continued search for Allied Chemical's damaging legacy, I had
two more stops to make. The first was over to Camillus, another village
just north of Solvay. Camillus, where one of the old wastebeds is located, became the favored site for sending the dredged material from the
lake bottom, much to the chagrin of area residents. The deposit site is
now covered up with topsoil and grass, but for a period of several years,
starting in 2012, it was open as the dredged material flowed into some
large porous plastic pipes, which allowed the water to drain out, leaving
the contaminated sediments. While this process was ongoing, residents
described an intense smell of mothballs. The New York Department of
Environmental Conservation downplayed any problems, in spite of residents' complaints that generated several lawsuits.[9]

The site sits on top of a hill near the old Erie Canal and, conveniently,
there's a bicycle path that runs along the canal leading part way to the
site. After walking along the trail for a bit, I scrambled up a fairly steep
hill, through some bramble, with the fenced off area of the site to my left.
At the top of the hill, there's a locked gate, with the site visible inside the
fencing. But there's not much to see—just a flat area covered with grass
and a pile of dirt sitting in the middle of it. The porous pipes that hold
the contaminated material are buried beneath, no longer off-gassing the
stew of toxics that drifted into the neighborhoods below, infused it with
the intense smells of body odor and mothballs. Fortunately, a dirt road
led from the gate back to the canal trail, so I didn't have to head back
down through the undergrowth.

After this, I made my way to the interstate and over to Jamesville,
which is southeast of Syracuse. Jamesville was the location of the quarry
that provided the limestone for the Solvay process. The quarry, now
owned by the large transnational firm, Hanson Aggregates, encompasses
2,238 acres. The company is a major cement manufacturer, and there's
now an operating cement plant at the quarry. The road leading up to the
site is private, so I couldn't get in to see it. I drove around through the
neighborhood to see if I could find a vantage point for viewing it, but
unsuccessfully. What I could see were piles of rubble lining the top of
the hillside, beneath which the quarry was located. There are number of
old, now abandoned, quarries in the area, which provided the limestone
for many of Syracuse's historic buildings.

From my investigations I did gain a sense of the scope of the environmental impacts of Allied's facilities. For many manufacturing processes, an industry's footprint can be spread over hundreds or even thousands of miles. With steel manufacturing in the U.S. in the twentieth century, iron ore was transported across the Great Lakes from the iron range of Minnesota. The coal to power the plants came from underground shafts in West Virginia or eastern Kentucky. The waste from steel making, and other forms of industrial production would be dumped into Lake Erie, where it would be dispersed in a large body of water, diluting some of it. The air pollution caused by burning coal generated intense localized problems, but it was also swept into the upper atmosphere, where it would rain down, slightly acidic, on the Adirondack forests and lakes or the Canadian Maritime Provinces. Nowadays, supply chains and their toxic impacts stretch out around the world. But with the Solvay process, things were more contained. The salt was taken from the Tully Valley and the limestone from Jamesville, so the legacies of environmental harms are local and tangible. The huge amounts of waste generated were deposited on land very close to the operations themselves, and then into a lake so small it was quickly overwhelmed. That is both the tragedy and value of Onondaga Lake. The concentrated and localized nature of these industrial activities made their ecological impacts highly visible and facilitated mobilizing against them. And since they were not, and could not, be dispersed widely, the difficulties involved with trying to reverse their impacts have become highly visible as well.

Environmental Histories

My investigation of Onondaga Lake uses the method of environmental history, which differs from conventional history in its focus. As Donald Hughes states in his *An Environmental History of the World*, "the task of environmental history is the study of human relationships through time with the natural communities of which they are a part, in order to explain the processes of change that affect those relationships . . . Environmental historians recognize the ways in which the living and non-living systems of the Earth have influenced human affairs." Moreover, "Like history itself, environmental history is a humanistic inquiry.

Environmental historians are interested in what people think about nature, and how they have expressed those ideas in folk religions, popular culture, literature and art."[10]

Environmental history is a relatively new field of inquiry, study, and analysis. Little that was written before the second half of the twentieth century could be considered specifically environmental history. As Andrew Isenberg notes (in 2014), "As recently as thirty years ago, textbooks in American, European world history contained little of what we would now call environmental history."[11] But the last few decades have seen a number of important works that have attempted an environmental history of large swaths of territory, such as North America, and others that have a more specific focus, such as the State of New York, or smaller locales, such as Love Canal. While these various studies are important and much welcomed, there's much more to do.

There are large gaps in our understanding of human impacts on environmental systems in North America and elsewhere. The United States has, for example, over 400 Superfund sites. A detailed accounting of the historical circumstances—the social and economic interactions that created these sites, the economic imperatives, the costs, and the benefits— has seldom been explored. At best, we often have surveys of toxic waste sites, which, while invaluable, do not provide exploration of the human and cultural contexts which brought them into existence, and which have attempted (or not) to address their continuing impacts. The global environmental system is vast, and opportunities for historical studies focused upon aspects of it are nearly limitless.

Some environmental historians emphasize the impacts of the material world on human societies, whereas others emphasize the role of human cultural systems on altering the environments that they inhabit. Ultimately, of course, the material and the cultural interact with one another in complex ways. The environments that humans inhabit influence the cultural systems that they create, which, in turn, alter environmental conditions, which then influence how cultures evolve, and so on. Still, no historian or historical analysis can do everything, so different emphases and starting points are perfectly legitimate, and all are useful.

William H. McNeill's work can be placed within the first category of how the material world shapes human society. *Plagues and People* is a benchmark in environmental history. Before McNeill, the impacts of parasites, fungi, microbes, and viruses on the development of human societies were given little, if any, consideration by historians. McNeill's sweeping history begins 100,000 years ago, focusing on human adaptation to microparasites, then shows how the Roman Empire was weakened by disease, considers the impact of the bubonic plague on European development, and demonstrates the important role that smallpox played in the colonial conquest of North Americas. McNeill's book, written in 1977, was highly influential, and could be considered among the first, if not the first, of comprehensive approaches to environmental history.[12]

McNeill's later writing became more expansive, and more explicitly environmental, when he published, with co-author Peter Engelke, *The Great Acceleration: An Environmental History of the Anthropocene since 1945*. The authors' conclusions are quite pessimistic. They see massive, irreversible, and humanity-threatening environmental transformations, without our having the political or imaginative capacity to address them. As they put it, "Since we cannot exit the Anthropocene, we will adjust to it, one way or another."[13]

Carolyn Merchant's work, to which I have long been indebted, provides a cultural approach to environmental history. She has explored how Indigenous agricultural practices helped to shape colonial agriculture, but also how European practices differed in important ways, especially in terms of the latter's demarcation of private property rights and patriarchal gender relations.[14] In *The Death of Nature*, she considered the role of the scientific revolution, especially the work of Francis Bacon, in marginalizing the knowledge and practices of women, who, before scientific rationalism excluded them, played a central role in mediating human relationships with the natural world, especially in terms of medicine and food production.[15] In *Reinventing Eden*, Merchant explored the cultural roots and ecological implications of European expansion into North America.[16] In *Earthcare*, she disinterred important unrecognized contributions of women (including indigenous women) to early American environmental organizing.[17] In her work, the exploration of cultural contexts is central to understanding environmental change, while recognizing the dialectical relationship between them.[18]

Other attempts to capture large trends in environmental history include Stephen Mosley's *The Environment in World History*, which covers broad topics: atmospheres, soil, forests, and cities, while also including case studies to illuminate human-environmental relationships in specific contexts. Local case studies provide important contributions with the frameworks established by the more general ones. In his words, the "micro" contributes to the "meso." He states, "Regional and local-level approaches, centered on different types of rural and urban 'ecosystems,'—coastal, forests, grasslands, riverine, market towns and industrial cities—can often produce more coherent case studies of how societies and environments shape and reshape each other over time."[19]

David Stradling's *The Nature of New York* is distinctive in its focus on a particular state. New York State, he tells us out at the outset, "abounds in places of national, even international significance." That is, of course, true, and likely true of every state. Stradling covers early settlement in New York, land clearing and environmentally devastating deforestation, agricultural practices, the construction of the Erie Canal, urban environments in New York City, the impacts of industrialization, and eventual implementation of conservation practices, with particular emphasis on the Adirondack Park, still the largest wilderness area east of the Mississippi River. But, while the breadth of the work is impressive, and provides a much-needed overview of a state with a long and rich environmental history, it has only a few sentences on Onondaga Lake in the context of chemical contamination threats in the state. This is understandable, but it does not provide readers with the necessary context to fully understand the complexity of the lake's contamination and cleanup.[20]

Numerous rich and vivid studies of particular environmental histories help to fill in gaps within the more global histories mentioned above. Larry G. Johnson, *Tar Creek: A History of the Quapaw Indians, the World's Largest Lead and Zinc Discovery, and the Tar Creek Superfund Site*, offers a kind of model for this study.[21] Johnson begins with the sojourn of the Quapaw from Ohio to Arkansas and an eventual forced removal to Oklahoma. He shows how the discovery of lead and zinc in the area that brought cooperation and conflict between the Quapaw and white mine-owners in the twentieth century. Johnson details the conditions of mine-workers, and the damage done to the health of workers, which helped spur the growth of the union movement, and he reveals the long-term devastating environmental impact of the mines. Johnson provides a means for understanding how the intersections of Native American histories and cultures, settlement, and displacement, are entwined with landscape. The specifics are unique, but the patterns of domination and devastation are not, and the events at Tar Creek provide lessons to understand the larger tableau of North American, and even global, environmental histories.

Another environmental history of a place is Joseph Alexiou's *Gowanus: Brooklyn's Curious Canal.*[22] Alexiou traces the history of Brooklyn's Gowanus Canal from the period of early Dutch settlement to the present. Gowanus Creek was transformed, through various fits and starts, from an estuary in the seventeenth century to a canal by the middle of the nineteenth. Alexiou details the parallel tracks that the creation of the canal took with the economic development of Brooklyn. Gowanus has certain parallels with the Onondaga Lake story, in that it, like Onondaga Lake, became a repository for a growing city's sewage. And, as with Onondaga Lake, the notoriously unpredictable precipitation events of the Northeast overwhelmed attempts to control sewage, turning the creek and then canal into a stinking, festering mess. And, as with Onondaga Lake, Gowanus was eventually designated as a Superfund site, and a cleanup plan was put forward that would involve dredging and a cap. But in the Gowanus case, some community members objected to the cleanup plan, because they wanted to keep the Superfund designation, which stymied commercial development and preserved the character of the community. In this way, the Gowanus case is unique.[23]

Another important work of local environmental history is *Remaking Boston*, edited by Anthony Penna and Conrad White. The chapters cover a wide range of topics. The first part of the books deals with the geological formation of Boston Harbor, the marshes that surround it, their interconnections with the land upon which the city was built, and the ecological relationships between the city and the countryside surrounding it, as the city developed and encroached upon ever-wider environments. We see the growth of industry, trade, and various forms of transportation, and how each of these sectors affected, not only natural areas, but the expanding human populations of the city and its suburbs. The final section of the collection focuses on climate and weather, and how the interactions of the city with other environmental elements, including the human, helped to shape and continue to shape the Boston Metro area.[24]

Martin Melosi's *The Sanitary City: Environmental Services in Urban America from Colonial Times to the Present*, rather than focusing on one place, focuses on one important, if generally unseen aspect, of nearly all American urban spaces, that is the treatment of waste. Melosi's study is of special importance to the current enterprise with its focus on sewage issues, a set of problems central to the Onondaga Lake story. Melosi is especially adept at revealing connections between political forces (machine politics and reformist politics in American cities), concepts of disease (the shift from the "miasma" to the "germ theory" of disease), and acceptance of engineering proposals that would mitigate sewage contamination in urban and suburban areas.

The specific character of much recent environmental history makes sense, given the emphasis of environmental thought and action on locality and place. As with the old adage, "think globally, act locally," environmental history has generated important large-scale, big-picture perspectives on the evolving shape of the global environment, while also focusing on the impacts of human activity on regions, states, cities, communities, and neighborhoods. There is a still a great deal of work to do to track all these things, but my hope is that *Toxic Lake* makes a contribution to ongoing scholarly activity.

Environmental Movements

There's truth to Philip Shabecoff's statement that, "The rise of environmentalism in the 20th century is likely to be remembered as one of the landmarks of human social development—a time when the effects of human activity on the natural world were not only broadly recognized but also acted on by peoples and governments."[25] But what is environmentalism? Is it a set of attitudes? A set of practices? A political movement or collection of movements? A main focus of Part II of *Toxic Lake* examines Onondaga Lake's contamination in terms of these questions, with the hope of casting some light on them.

The conventional depiction of the history of environmentalism in the U.S. describes the late nineteenth and early twentieth centuries as the era of conservation: the publication of Henry David Thoreau's *Walden*, the writings of Ralph Waldo Emerson, the poetry of Walt Whitman, the creation of the national park and federal lands systems, the conflict between Sierra Club founder John Muir and Interior Secretary Gifford Pinchot on the meaning of conservation. In the conventional narrative, the expansion of environmentalism to question industrial society itself, what is sometimes dubbed as the "modern" environmental movement, was launched with the publication of Rachel Carson's *Silent Spring*, a book which criticized the use of pesticides, specifically DDT, but also questioned the impacts of the larger systems of industrial production that its use represented, on human health and the natural world.[26]

As the environmental movement expanded, so the narrative proceeds, it became more influential, leading to the passage of meaningful environmental legislation: the Clean Air Act, the Clean Water Act, the National Environmental Policy Act, the establishment of the Environmental Protection Agency, and the Endangered Species Act. In the 1970s, the discovery of Hooker Chemical's contamination of Love Canal focused attention on the very ground on which homes and cities are built, and the potentially health-threatening chemicals that they contain.[27] The result was passage of the Comprehensive Environmental Response, Compensation, and Liability Act of 1980 (CERCLA) or the Superfund Act. Continuing onward, the 1980s saw the rise of the environmental justice movement, which had its start in Warren County, North Carolina, when local residents attempted to block the siting of a toxic waste disposal site.[28] Scholars, Robert Bullard, being primary among them, demonstrated racial disparities in the siting of waste disposal sites, especially in the southern United States.[29]

The simplicity of this narrative is satisfying, and it holds some truth, at least in general terms. But as with all grand narratives, it is both incomplete and partly inaccurate. Numerous scholars have added to the narrative and questioned the singular primacy given to key figures. Carolyn Merchant, for example, has noted John Muir's racial insensitivity, if not outright racism, and highlighted the wilderness movement's supersession of Native American land claims.[30] Kimberly Smith has revealed the roots of African American environmentalism in the antebellum South and its continuity and evolution through the nineteenth and twentieth centuries.[31] Merchant has also documented the contributions of the women's suffrage movement to early twentieth-century conservation.[32]

Within the larger narrative, Rachel Carson is often giving credit for inspiring the modern environmental movement. Linda Lear's comment in the Introduction to the 2002 edition of *Silent Spring* is not untypical. Carson, she stated, "set in motion a course of events that would result in a ban on the domestic production of DDT and the creation of a grassroots movement demanding protection of the environment through state and federal regulation. Carson's writing initiated a transformation in the relationship between humans and the natural world and stirred an awakening of public environmental consciousness."[33] But while Rachel Carson's contributions should not be discounted, given the public attention and the warranted acclaim that *Silent Spring* received, including an endorsement by President John F. Kennedy, noteworthy environmental catastrophes had occurred before *Silent Spring*'s publication, garnered public attention, and pointed to the health and ecological dangers posed by industrial production.

The Donora, Pennsylvania, disaster of 1948 is among the most well-known. Twenty people died when a temperature inversion trapped the smoke from a steel plant in the Monongahela Valley. The stagnant atmosphere contained mixture of fog and fluoride effluent from a Carnegie zinc smelting plant. While the event itself received national attention, less well-known is the environmental organizing that preceded it. The facility's siting in the valley generated considerable resistance from local farmers earlier in the century. In the 1930s, residents brought unsuccessful lawsuits against the Carnegie corporation in an attempt to gain compensation for damages done to their property and livelihoods. As was typical of such instances of industrial development, however, as the complex expanded, the power of the company and its influence on local officials, along with economic interests of the 14,000 employees, overwhelmed doubts about the environmental costs.[34] There are parallels here to the Onondaga Lake case, specifically in terms of the company's power to suppress doubts about the environmental impacts of its operations. Likely there are numerous other cases of resistance to harms associated with industrial development early in the twentieth century, but they become lost within the broader arc of the conventional narrative.

Spears has noted that attention to DDT preceded the publication of Carson's book. Two sisters, Dottie Colson and Mamie Ella Plyler, raised questions about DDT's impacts as early as 1945, when children complained of having their throats burned from milk that was produced near Camp Stewart, Georgia, an area that had been heavily treated with

DDT. They were, of course, ignored. But national press did give attention to DDT in the 1940s, and international conferences considered the impacts of poisonous chemicals.[35]

Chad Montrie's *The Myth of* Silent Spring, as the title suggests, goes even farther than Spears in questioning Carson's centrality to the launching of modern environmentalism. Montrie suggests that Carson placed too much emphasis on the contemporary threats of her era, and thus erased previous opposition to the impacts of industrial processes. He states, "by opening *Silent Spring* the way she did, casting ruinous environmental change as a uniquely twentieth century phenomenon, Carson misinterpreted history, relying on an inventive portrait of the past to make her jeremiad seem even more timely and necessary." Whereas, in fact, "the particular problems that gave rise to the modern environmental movement began more than a century earlier (at least), with the onset of industrialization and its many attendant consequences."[36] Her focus, was, in other words, too narrow. It ignored nineteenth- and twentieth-century sanitation movements, labor organizing against the impacts of pollution on workplace health and safety, and the myriad grassroots efforts that attempted to protect their communities from the incursions of industrial production.

A singular focus on one event, person, or book minimizes the larger historical contexts at play as environmentalism has evolved. Post-1945 environmental movements appeared within the atmosphere generated by adoption of the U.N. Universal Declaration of Human Rights, which referenced health and well-being as human rights. Moreover, "the rate and scale of environmental change in the postwar period—the Great Acceleration—was unparalleled." Consumerism, and the production that made it possible, exploded. And as this occurred, its impacts became more visible and potentially alarming. The wealth of large corporations expanded dramatically in the post-World War II period, encouraging suspicion of their power over economic and political life. Cheap energy became widely available, along with the wastes that it produced. Automobility became the dominant form of transportation, which generated atmospheric pollution. Atomic weaponry threatened the existence of human life. Agriculture became more dependent on chemical and seeds companies. Synthetic chemical production expanded on an exponential curve.[37]

What has emerged from various efforts to rethink the environmentalist narrative is a rich, complex history that does not deny the broad parameters of the conventional narrative but provides more detailed and nuanced enhancements of it. Lost histories of environmentalism focus on local struggles that have been left out of the larger narrative and do not always fit comfortably within it. Participants in local struggles are often conflicted, torn between their direct economic interests and tangible concerns about health, safety, and ecological protection. They may or may not see themselves as connected to a larger movement of environmental progress (or decline). Gould et al., for example, focus on "citizen workers," who are concerned with local health and safety issues, such as problems generated by toxic contamination, but who also gain benefits from systems of production that create the contamination. They fear the loss of property values if industries shut down or flee. They are caught in the middle, in other words, between supporting local economic activity that may undermine ecological systems and recognizing its negative externalities.[38] The history of American environmentalism is filled with important contributions such as these, and attention to local circumstances, through the detailing of case studies, helps to fill in these gaps.

In recent years, there has been increased attention to local environmental actions. Carol Hager and Mary Alice Haddad ask readers to rethink the concept of NIMBY, or "not in my backyard," and reconsider NIMBYism as a positive form of environmental actions, drawing from case studies around the globe.[39] Nikolay Mihaylov and Douglas Perkins draw upon psychology and social movement theory in an attempt to examine the conditions within which local environmental organizing is spawned. As they note, "Local environmental grassroots activism is robust and globally ubiquitous despite the ebbs and flows of the general environmental movement."[40] The case studies that are being initiated now are global in scope, but local in their focus. Robert Vickers has examined local environmental activism in Britain.[41] Allen Hassani-yan has studied local environmental organizing in Kurdistan, drawing upon the concept of "subaltern environmentalism."[42] The journal *Local Environment*, which was launched in 1996, is replete with attention to local efforts operating worldwide. Examples of recent studies include environmental governance and mining in northern Ghana, organizing to overcome water poverty in Barcelona, potential brownfield development in Osaka, Japan, and local indigenous climate practices in Bangladesh.[43]

Environmental organizing around the pollution of Onondaga Lake unfolded over decades, spawning a complex and varied range of participants over time. It involved middle-class, often white, citizen activists, nonprofit organizations, environmental scientists, public officials, public interest law firms, members of the Onondaga Nation, and residents of an historically significant and marginalized African American community. As such, it cannot be captured under the umbrella of one specific form of environmental organizing, such as "mainstream" or even environmental justice. Yet while Onondaga Lake environmental activism created its own set of internal dynamics, it took place within a context provided by larger movements and, as such, can be understood via reference to them.

In the postwar era, an expanding middle class of affluent property owners exercised their economic and political power by challenging industrial pollution. These national trends were evident in opposition to Allied's practices. Environmental science became increasingly linked with environmentalism as the complexities of addressing ecological problems became more obvious. Environmental scientists from Syracuse University and New York State's School of Environmental Science and Forestry, and local activists trained who worked with them, took a lead role in investigating Onondaga Lake's contamination and proposing remedies.

Public officials were both pressured by local citizens and empowered by environmental legislation that was passed at the end of the 1960s, nationally and in New York State. Senator Daniel Patrick Moynihan held a set of public hearings on Onondaga Lake pollution that placed it within larger national debates. Public interest law firms, which began to appear in the late 1960s, often inspired by the work of Ralph Nader, arose to use the new legal frameworks to defend environmental rights.[44] The Atlantic States Legal Foundation brought a major lawsuit against Onondaga County to improve its sewage disposal practices. But to comply with the lawsuit, a federal judge approved a plan that would place a satellite sewage treatment facility in an African American neighborhood, against the wishes of residents, as well as the Atlantic States Legal Foundation. The Onondaga Nation's assertion of land claims related to the lake's despoilation took place within a wider contest of Native American land claims in New York State, but it was unsuccessful.

The Onondaga Lake case reveals a complexity of competing ideas and interests. Fractures appeared between scientists and public officials, citizens groups and members of the Onondaga Nation.

* * *

What follows is a study that traces the environmental harms done to Onondaga Lake and the attempts at remediation. I have organized the book into three parts in order to provide a conceptual framework for the chapters within each. Part I shows Onondaga Lake in the context of local and global environmental histories and includes chapters 1 and 2. Chapter 1 provides cultural and geological background, the historical context for what would eventually shape Onondaga Lake's environmental future. In this chapter, I consider the Indigenous cosmologies that fostered respect for the earth and appreciation of its bounty and discouraged practices that brought damage to its ecological systems. I examine Haudenosaunee narratives of Peacemaker and Hiawatha, who together formulated the Great Peace and ended an intense period of conflict between the Five Nations to establish the Haudenosaunee Confederation. I then trace the displacement of Indigenous worldviews by European scientific ideologies, which provided the cultural frameworks and technical means to exploit the Onondaga watershed, and which laid the foundations for the eventual siting of the Solvay facility.

In chapter 2, I discuss the history of soda ash production through the nineteenth century. Soda ash was a highly prized chemical substance, crucial for glassmaking and other manufacturing. The French, known for their glassmaking skills, were a major market for soda ash, which they were able to obtain from Scottish farmers who refined it by burning kelp they gathered from shorelines. When war with England disrupted these markets, King Louis XVI offered a prize to anyone who could develop an artificial process for soda ash production. Nicolas Leblanc invented the first, which worked enough to generate soda ash but was complex and highly toxic. Leblanc facilities spread across Europe, but the process's limitations were obvious. Others attempted to find a simpler, cheaper, less toxic method for creating soda ash.

The Belgian chemist Ernest Solvay, after many false starts, discovered a workable and economical replacement for the Leblanc system by the end of the nineteenth century. Recognizing that the Syracuse area provided the necessary constituents for it, primarily salt and limestone, William B. Cogswell convinced Solvay to allow him to open a franchise for the new process. The result was a highly successful manufacturing hub that became a magnet for immigrants from the Italian Tyrol who provided labor for the Solvay facility and established a stable and prosperous community. But the Cogswell plant and additions to it generated much of the waste that contaminated Onondaga Lake and the land surrounding it.

The second part of the book focuses on environmental movements. In chapter 3, I document the symbiotic relationship that developed between Solvay process wastes and Onondaga County sewage treatment. Decisions made by the City of Syracuse to combine storm and sewage runoff had long-term and deleterious consequences. The city also made an early decision to build a cheap and minimally effective sewage disposal plant that contributed significantly to pollution in Onondaga Lake that lasted for decades. Ironically, however, city and county officials encouraged Allied Chemical to dump calcium chloride wastes on the land and into the water, as a way to control the smells that resulted from the disposal of raw sewage into the lake. The calcium chloride reacted with the phosphorous, sulfur, and potassium in the sewage to reduce its smell and disguise the extent of its impacts. Some Allied officials suggested that the Solvay process wastes "cleaned" the lake. The company and county officials struck a devil's bargain that allowed each entity to disown responsibility for the environmental damage they were doing.

In chapter 3, I also examine some of the early, rudimentary attempts by community members to challenge Allied Chemical, in order to halt the ongoing pollution of the lake. Opposition to Onondaga Lake contamination represents an early local attempt at environmental organizing that occurred more than a decade before a national environmental movement gained prominence in the 1960s. In fact, environmental organizing related to Onondaga Lake, along with other local actions across the U.S., undercuts the conventional narrative that the environmental movement began in the 1960s, after the publication of Rachel Carson's

Silent Spring. The Onondaga Lake story provides a microcosm of how American environmental awareness evolved in the 1940s and 50s.

In chapter 4, I track how, in the 1960s and 70s, as the environmental movement began to take hold nationally, demands for Onondaga Lake cleanup continued to grow, and conflicts between community groups and the company intensified. Environmental activists who focused on Onondaga Lake helped to foster an environmental consciousness in New York State and even beyond. Over time, public officials at local, state, and national levels were forced to address the contamination issues that Onondaga Lake represented.

I also follow the closing of the Solvay plant and the initiation of serious attempts to clean up Onondaga Lake. The New York State Department of Environmental Conservation brought a lawsuit against Allied to address the lake's environmental problems. The closing of the plant had a negative economic impact on the small Village of Solvay, which lost its main employer, but it also offered hope that the lake could be restored. Sadly, as the lake became the focus of intense scientific study, the extent of the damage to it became clearer, alarmingly so, when multiple complex intertwined forms of contamination were revealed. A large pile of what was first described as "goop" was discovered in the lake bed. It turned out to be an amalgam of calcium chloride, toxic organic chemicals, and large amounts of mercury. Dealing with the goop would pose one of the most intractable problems for a lake cleanup.

The focus of part three is environmental restoration. In chapter 5, I trace the cleanup efforts as they moved forward into the twenty-first century. When AlliedSignal bought Honeywell, to help shed its reputation as a polluting chemical company, relations with the New York Department of Environmental Conservation improved. The company reached a compromise agreement with New York State to commit to a one-half-billion-dollar remediation effort. Honeywell had first proposed fairly minimal dredging and other restoration actions, whereas the state, initially at least, wanted virtually the entire lake bottom dredged. The two parties arrived at a compromise, where some dredging and capping would occur, and remediation efforts would also be initiated to address many of the toxic legacies that existed in the land surrounding the lake. Importantly, Onondaga County constructed a state-of-the-art tertiary treatment facility to address the sewage issues. However, not all parties were happy with the final result. The Onondaga Nation launched legal actions, ultimately unsuccessful, to require the state and Honeywell to do a more extensive cleanup. The state and many local environmental activists seemed to hold a narrower and more utilitarian view of the lake's value than did the Onondagas. Their differences can be traced back to philosophies and priorities outlined in the book's earlier chapters.

In chapter 6, I examine the ongoing efforts to address the multiple contamination problems that undermined the lake's ecological recovery. These included capping and dredging operations, reestablishing flora on abandoned wastebeds, and filtering contaminated groundwater. Unexpected complexities emerged. For instance, the drop in the lake's nitrogen content, resulting from improvements in the Metro Sewage Treatment plant, had the indirect effect of increasing methylmercury levels. As a result, researchers developed a plan to inject nitrates into the lake bottom to reduce its release (a practice that is ongoing). Disposal of the dredged material from the lake bottom provoked local opposition from residents of the suburban community near which it was deposited.

The legacy of salt removal via water injection, was causing subsidence in the hills in the Tully Lake area, ten miles south of Syracuse. It created mudboils, as rainwater poured through empty caverns and shot up into Onondaga Creek, in geyser-like fashion, filling the creek with mud, and making is useless for fishing or swimming as it ran through Onondaga Nation territory.

Due to a lawsuit by a local environmental law firm, Onondaga County was now under court order to curtail overflows into the Metro treatment plant. In response, it proposed constructing several interceptor facilities, to capture and moderate them during severe runoff events. The first was to be located in an historically Black neighborhood and would require demolition of several houses. Community opposition developed widespread support, and the county eventually agreed to implement the Save the Rain program, which used decentralized green techniques, such as tree plant and porous pavements, as an alternative means to staunch overflows, but not before the first facility was constructed. Save the Rain, which gained widespread community support, established Syracuse as a city on the forefront of the urban greening movement.

Finally, I look at recent developments, including the declaration by the State of New York that the Onondaga Lake cleanup has been completed. What that means is a matter of some dispute. I explore the meaning of environmental restoration both as a concept and a set of actions. Onondaga Lake has clearly been improved over the last two decades, but it is still a Superfund site, due to the buried sediments at the lake bottom. And it is unclear if or when the Onondaga Nation's vision of a sacred lake will ever be reestablished.

Whatever the ultimate ecological fate of the lake, the process by which the cleanup effort was determined was clearly flawed in that it continued the marginalization of the voices of the Onondaga Nation, an historical injustice that can be traced back to the period of colonization. One possible strategy going forward is to follow Canada's lead and include Indigenous knowledge as a requirement for environmental policy-making.

PART I

Environmental Histories

1

Origins

Indigenous Cosmologies and the Science of Geology

It might seem odd, or as an unnecessary distraction, to begin a work, the focus of which is environmental contamination and attempts at restoration of Onondaga Lake, with a discussion of Indigenous cosmologies and the origins of the science of geology. But, as a patient reader will learn, these competing worldviews collide in very tangible policies and centered disputes regarding the current ecological health of Onondaga Lake and what should be done to address the harms that have been committed against it.

Onondaga Lake represents a microcosm of the conflicts and contradictions of two competing worldviews, one indigenous to the region, with roots, traditions, and practices stretching back thousands of years, the other transplanted from Europe by colonists who imposed their traditions, epistemologies, and cultural practices on the people they encountered. These are not long-dead struggles but living conflicts with relevance to contemporary environmental politics and policies. In the case of Onondaga Lake, the struggle between Indigenous knowledge and practices and the decisions of corporate managers and public officials in New York State came into focus through attempts to address the lake's pollution problems. At the heart of this conflict was how the lake's significance is conceived differently by the Haudenosaunee, who embrace both its practical and spiritual value, and European immigrants and their descendants, who viewed it in narrower, primarily instrumental terms.

Cosmologies represent a group's particular ideas about the order and structures of the universe. They matter because they provide the cultural backdrop, the parameters from within which the cultural practices emerge. The complexities of Haudenosaunee cosmology are staggering. Forty different versions have been identified, derived from oral traditions that varied from community to community and evolved over time. Written versions were transcribed and translated by European colonizers or non-Native Americans, who brought their own perspectives and biases to their work. Thus, it is important to tread carefully and respectfully when attempting to interpret them, but to ignore these ideas in any serious work of American environmental history would be irresponsible.

John Napoleon Brinton Hewitt provides a starting point. His mother was Tuscarora. He was born on the Tuscarora reservation and raised by a Tuscarora family, members of the large and diverse Haudenosaunee community. He was also an ethnologist who worked for the Smithsonian Institution's Bureau of Ethnology from 1888 until his death in 1937. Hewitt's approach to documenting Haudenosaunee narratives could best be described as a form of transcription, that is, listening to and writing down the testimony of his interviewees. As a result, his versions are considered to be among the most accurate of any that exist in writing. Yet, given that Hewitt's versions were recorded at a relatively late date, we cannot know how faithfully they render the events of the preceding 300 years. Still, for all their potential flaws, the narratives provide a glimpse into the cultural context that preceded European colonization and all that followed. The Onondaga creation story is the longest and most elaborate of those collected by Hewitt in his monograph *Iroquoian Cosmology*. The recounting of that narrative that follows here is brief, compressed, and doubtless woefully inadequate, but it provides an essential context for understanding the political and legal controversies that would eventually engulf Onondaga Lake.

Onondaga Lake

In Sky World man-beings inhabited longhouses. Among them were a man and woman who lived on different sides of one longhouse in particular. The woman tended to the man's hair each day, until one day she became pregnant. When the man asked her who the father was, she refused to tell him. Soon he became sick and died. Death had not previously existed in Sky World. As the man was dying, he asked the woman to place his body up in a "high place" in a "burial case." The woman was grief-stricken, as were other members of the village. Still, they followed the man's wishes.[1]

Soon the woman gave birth to a girl, who grew "rapidly in size," and began to "run about." But she also began to weep, and her weeping was inconsolable. Members of the lodge carried her to see the corpse of the man who died, and she stopped weeping. As soon as she was taken down, she began to cry again. The villagers brought her up and down from the corpse, to keep her from weeping, until she was large enough to climb for herself. The corpse eventually told her that he was her father, and he gave her gifts, including a bracelet and a necklace.[2]

The corpse-man then told the young woman that she needed to marry, but that it would be a difficult and dangerous journey to meet the man she would marry. She would have to cross a river. She would then come to a lodge, in front of which grew a tree called "Tooth." Inside she would find the man she was to marry. They would spend the night together, and the next day, he would teach her how to make corn mush. She would do this without clothing, and even though the hot mush might burn her skin, she must continue with it. Moreover, the man's dogs would lick the splattered mush from her skin, and, this would be painful, breaking her skin. After she had completed the ritual, the man and woman would marry.

So, the maiden left and followed the corpse's directions. She spent two nights with the man-being in the lodge, known as the Ancient. He gave her a basket. He filled it with meat, shaking it seven times as he did so, and told her to return home and not to stop along the way, even if tempted to do so. Once home, she would tell other members of the community to remove the roofs from their huts, because that which is called "corn" would rain down upon them and that it is what man-beings would live on in the future.[3]

The maiden followed the Ancient's direction, meeting, along the way, Aurora Borealis and White Fire Dragon, who asked her to stop. She ignored them. Once home, she told the villagers to remove the roofs from their lodges and went to see her father, the man-corpse and told him that she had done as directed. He expressed satisfaction and told her that the man in the lodge was He-Who-Holds-the-Earth. Then the corn began to rain down on the villagers, who went to sleep, waking up the next day to find their lodges filled with it. And their chief told them to appreciate the gift they had been given by He-Who-Has-the-Standing-Tree-Called-Tooth.[4]

The maiden then returned to He-Who-Has-the-Standing-Tree-Called-Tooth. A change came over her. She became pregnant because she and her betrothed "breathed together." Henceforth, this would be the manner by which man-beings, as well as animals would produce offspring. But her betrothed, now referred to as the chief, was confused because they had "never shared the same mat." When he asked her who it was that made her pregnant, "She did not understand the meaning of what he said."[5]

Soon, as the maiden's pregnancy became more apparent, He-Who-Has-the-Standing-Tree-Called-Tooth became ill. He told her that she would give birth to Zephyr, which confused her, but nevertheless, she gave birth. She lay the child on a mat for ten days, and then took it away. The chief's illness continued. He told the villagers that they must pull up the tree called Tooth, tearing the earth open, and afterward lay him down near the "abyss" created with this wife nearby. Chief then told the maiden to sit with him on its edge, and, when she did so, he pushed her into it and declared himself well again.

The maiden fell. She fell into water, deeper and deeper, and neither Loon nor Otter could stop her. Finally, the Great Turtle succeeded in stopping her, and all the animals began to place earth on its back until it grew and became the Earth. The next day the woman saw a deer lying nearby, and she roasted its meat with wood given to her by the chief. And then she gave birth to a female child and erected a thatched hut where the two of them resided. One day a man-being came to talk to the younger woman, spent the night, and soon she was pregnant with two male man-beings.[6]

The twins argued about who would be born first. One was ugly, covered with warts, the other was handsome. The handsome one arrived first in the normal way, but the wart-covered one came out through the maiden's armpit, killing her in the process. When the grandmother asked who had killed the maiden, the ugly one (Flint) blamed it on his brother (Sapling). The grandmother, believing Sapling to be a liar, threw him out of the village, where he landed in grass. She attended to Flint, who grew rapidly.[7]

Sapling traveled over the earth and observed everything that was there. He made the bodies of all the earth's animals, all in pairs, males and females. These included birds, who when gathered made terrifying sounds, and the game animals, such as deer. These were all for the happiness of the man-beings who would soon dwell on Earth.[8] But the next time Sapling left on his travels, Flint put all the animals into a cave and placed a stone over the entrance. Upon his return, Sapling let the animals out of the cave, but Flint was able to return to the cave entrance after Sapling left, and again placed the stone over it, trapping some animals inside. These trapped animals would not escape during the time of man-beings and would continue to haunt the earth.[9]

Sapling then encountered Hadu, who also claimed mastery of the earth. Sapling challenged him on his claim and told him to move a mountain if he was really the master. When Hadu was unable to accomplish this, he in turn challenged Sapling. Sapling moved a mountain directly up behind him, and Hadu conceded Sapling's superior claim and pledged his assistance. Hadu insisted that he would aid the man-beings when they came to dwell upon the earth, and that they could receive his assistance when they fell ill from various infirmities that might affect them, even though these infirmities that Hadu was responsible for creating. He was "grandfather," and the man-beings were his "grandchildren."[10]

Sapling again traveled the earth, returning to his home, where he found grandmother. She took the head of his dead mother and fashioned it into the sun and fashioned her body into the moon. Sapling pursued the sun and asked for help to obtain it, which Fish, Beaver, and Fox agreed to. They needed a canoe, so Beaver helped to build it. When morning arrived, Sapling said that he would unfasten the orb from a treetop, and Fish and Fox agreed to help. Fish passed the orb to Fox and the two fled with it. Grandmother was enraged, and chased after Fox, Otter, and Beaver in pursuit of the orb. She called out to each in an attempt to stop them from moving the sun "hither and thither," but her power could not influence Fox or Beaver. Otter relented, however, and when he stopped, Beaver hit him in the face with the canoe paddle, flattening it. Now they could return home with the sun, and Sapling was able to fix it up in the sky so that man-beings could have light.[11]

The final scene in Hewitt's version of the story seems to have been influenced by Christian missionaries, since it tracks in some ways with Genesis. Here Sapling goes to the river, where he fashioned the body of a man-being and then the body of a woman-being, and they became married. But they were not interested in one another, so when they slept, Sapling exchanged their ribs. When they awoke, he told them to care for one another, and that they would populate the entire earth with their offspring. But Sapling noticed his brother Flint working at creating his own set of man-beings. Sapling told Flint to stop, because he did not have the power to create true man-beings, and instead, Flint created apes, monkeys, and horned owls. When Sapling told him to stop, he refused. Sapling cast him down into the earth where it was very hot. At that Flint asked to talk with Sapling, who agreed. When they spoke at the designated meeting place where "the earth is divided in two," Flint was unrepentant. So, despite his plea, Sapling threw Flint back into the place where the earth was hot, where he would remain until the end of the earth.[12]

Sapling, however, would grow old and then "recover his youth" in a constant process of aging and regeneration. He was "possessed of force and potency," which would be forever "full, undiminished and all sufficient." And this vigor would be true of all the earth's beings, which would be "affected in the same manner, that they severally transform their bodies, and, in the next place, that they retransform their bodies, severally, without cessation."[13]

In analyzing this narrative, we should note that it does not start with a void or nothingness, nor is there a beginning *per se*, but rather a pre-existing Sky World the origins of which seem to be outside the horizons of human understanding. Humans can't know everything, and it is unimportant that they do. Life in Sky World is, in many respects, similar to what would become life on Earth. In Sky World, people live in villages at the center of which are longhouses. Perhaps Sky World could have continued indefinitely, but a break occurs, and this sets off a chain of events that ultimately ends in the creation of earthly man-beings. It is by no means a straight path, however. The break in Sky World occurs when a maiden becomes mysteriously pregnant. She doesn't know or won't admit to knowing how it occurred or who impregnated her. But the pregnancy causes her mate to become sick—perhaps with jealousy—and die. Mortality enters into the world. Birth yields the inevitability of death.

Yet when the first man-being dies, he doesn't disappear. Rather, he continues to communicate and influence the community in significant ways. The dead are always with us, and we would do well to listen to them. He directs his daughter to undergo a dangerous task for the good of the community, which yields a great gift: corn. The story speaks to women's strength and their central role in agricultural cultivation in a matrilineal society. But the community is composed not only of women and men, but of otters, bears, and wind currents. All are woven together in the fabric of life in Sky World, and these relationships are replicated on Earth.

Once on the earth, which has been created by animal spirits, and to which she has been transported safely by them, the maiden generates the flora that carpet the earth. She gives birth to a child who has great power, and who is in turn impregnated by a handsome, nameless man who disappears, leaving her alone to give birth to the twins, Flint and Sapling. In birthing them, she sacrifices her own life. The woman's sacrifice generates the earth's fecundity as well as its instability. The two are inseparable. The entry of the twins, Flint and Sapling, brings conflict, violence, instability, competition, and change. The world is not static, and the twins encapsulate its dynamism and shape-shifting. Flint isn't evil, in a Christian sense, but he is a troublemaker. His mischief has a purpose. He makes life difficult for man-beings, and so he tests and challenges them, making them stronger, more resilient, and smarter. Without the Twins, Earth would be just another Sky World, and mortality would have been introduced into it, disrupting it, for no reason.

Important to note is what the earth isn't. It's not fallen. It is not a desert. It didn't originally contain a perfect, bountiful garden. And neither men nor women are given authority or mastery over it. In other words, it is not the world of the Old Testament. The *Encyclopedia of the Haudenosaunee* states that the creation story

> deals with the positive spirit energy of *uki* or *ugi* (called *orenda* by J.N.B Hewitt), which is balanced out by the negative energy of *otgont* or *otkon*. These spirit powers should be recognized as bonded pairs. They should not be separated from each other's influence and certainly not be confused with any sort of Manichean dichotomy. Creation involved accident, recklessness, and madness, but no satanic evil; there was also forethought, responsibility, and clarity, but no Jehovan godliness. Nothing that was done could not be undone. Everything was mediated. Creation was a cooperative effort of many spirits.[14]

William N. Fenton distills a number of "principles" from his various versions of this origin story, among the most pertinent for our purposes are these:

> The earth our mother, is living and expanding constantly, imparting its life-giving force to all growing things upon which our lives depend.

> The alteration of seasons and tasks is attuned to ecological time as are the lives of plants and animals that rest when the earth sleeps.

> A chain of kinship connects all members of society from the dead to the smallest child whose life is separated from the spirit world as the thinness of a maple leaf.

> *Do not oppose the forces of nature* . . . *Orenda* (supernatural power) adheres to animate and inanimate things, to aspects of the environment, and to sequences of behavior . . . [or] *kanyonyonk*.[15]

Keep in mind that the strength of Native American societies lies in their diversity and adaptation of narratives to local cultural and material circumstances. As Lee Irwin notes in his sweeping analysis of Native American religious traditions, "Among precontact Native religions, there was no 'one way,' only a multitude of paths, each unique to its own community and each embedded as a complete and complex way of life."[16] Every interpretation of this material must be considered within this context.

These cultural traditions aren't simply a matter of historical interest. They are alive and practiced by contemporary members of the Haudenosaunee Confederacy. Clan Mother Wendy Gonyea explained to me that before each ceremony or gathering of the Onondaga community, they give thanks. They thank the grass, the animals, and the water: "the springs, the small puddles, the ponds, the streams, and the running waters. Water is a gift." Such recognitions are central to Onondaga cultural practices. The contamination of Onondaga Lake, which will be discussed in much greater detail going forward, has been, given these traditions and practices, an especially distressing situation. It is not only an affliction of the environment; it is a deep violation of a culture.

Onondaga Lake in Indigenous and Settler Histories and Cultures

The Onondaga Lake watershed and Onondaga Lake itself are of particular cultural importance for the Haudenosaunee. What is now New York State was land that they occupied for thousands of years before the arrival of Europeans. It is also a place in which the dramatic story of the fashioning of the Great Peace transpired. The narrative of the Great Peace is associated with Hiawatha, a name familiar to most Americans, largely due to the poem *The Song of Hiawatha*, published by Henry Wadsworth Longfellow in 1855. Longfellow's appropriation of Hiawatha's name for nominally aesthetic, but actually instrumental, purposes is a corruption that in symbolic terms mirrors the contamination of Onondaga Lake itself.

In Longfellow's hands, Hiawatha became a noble and romantic savage, a man who invented picture writing, fasted for his people, fell in love with and married Minnehaha, and lost her to a terrible famine. Longfellow's Hiawatha lived in a wigwam, a form of habitation used primarily by First Nations people in Canada, such as the Ojibwe and Innu, neither of whom was associated with the Haudenosaunee; nor had either been members of the Haudenosaunee Confederacy. The poem was wildly popular during its time, and a great commercial success. The tone now might seem patronizing, but its success is understandable. The narrator directs it to all those "Who believe, that in all ages / Every heart is human, / That in even savage bosoms / There are longings, yearnings, strivings / For the good they comprehend not."[17] True, the poem references the catastrophe prefigured by the appearance of Europeans. Iago and Hiawatha relate visions of the "white-man," in which they "beheld a nation scattered, / All forgetful of my counsels, / Weakened, warring with each other."[18] But resolution in the poem occurs when Hiawatha receives priests into his wigwam and converts to Christianity. In the final scene, Hiawatha enters his canoe, says farewell, and paddles toward a paradise that is a heavenly synthesis of Indigenous and Christian conceptions.[19]

Longfellow borrowed much of the narrative of Hiawatha from American storyteller Henry R. Schoolcraft, who, like Longfellow, lived and wrote in the first half of the nineteenth century. Originally from Albany, he claimed an anthropologist's interest in Native American cultures, and married an Ojibwe woman, Ozhaguscodaywayquay, who took the Anglo name Susan Johnston. He had an interest in exploration, geology, and mining. He traveled through Illinois and Missouri, attempting to locate areas where mineral deposits—primarily lead—might be found and purportedly "discovered" the source of the Mississippi River. He spent a good deal of time living with Native people in the Midwest, and eventually published a multivolume reference work on American Indians.

Schoolcraft has been variously praised and criticized for his associations with and writing about Native Americans. Longfellow cribbed his Hiawatha story from an essay that Schoolcraft had written. Longfellow also took liberally from the *Kalevala*, the epic poem of the Finnish nation, compiled by Elias Lönnrot from oral traditions and first published in 1835.[20] According to the introduction of a 1968 edition of *Hiawatha*, this narrative "may well have reminded him also of the Indian legends, which have that likeness to the Finnish that springs from a common intellectual stage of development and a general community of habits and occupations."[21] Such comparisons to stages of development are simplistic and false, but they may have existed in Longfellow's thinking, and he borrowed from whatever sources served his poetic interests. Unfortunately for historical accuracy, Schoolcraft's (and Longfellow's) Hiawatha refers to a legendary figure of the Ojibwe, not the historical person Hiawatha. Longfellow, according to his notes, considered using the original name for this figure, Manabozho, but dropped the idea, because he didn't like the way it sounded.[22] As a result, confusion has followed the historical figure of Hiawatha ever since and has obscured his important contributions to the history of the Haudenosaunee and of the North American continent.

Bruce Johansen, writing in the *Encyclopedia of the Haudenosaunee*, is scathing in his appraisal of Schoolcraft's version of the Hiawatha story:

> Schoolcraft's Hiawatha bore no resemblance to the historical figure cherished in Haudenosaunee tradition. Schoolcraft turned him into an Annissinabe (Chippewa) and confused him with the Annissinabe (Ojibway, also known as Chippewa) cultural hero Nanapush ("Manabozho"). Ignorance was not to blame here, but a raging disrespect for Native cultures: Schoolcraft knew the difference, but simply liked the sound of "Manabozho." In addition—and this probably was ignorance—

Schoolcraft confused Hiawatha with Tarachiawagon, one name for the Peacemaker. Finally, Schoolcraft plagiarized Joshua Clark's *Onondaga, or, Reminiscences of Earlier and Later Times* (1849), pretending that the research was his own.[23]

One of Schoolcraft's most glaring errors, even if not deliberate, was the conflation of Hiawatha with Tarenyawagon, the Peacemaker, the historical figure whose vision brought law to the Five Nations of the Haudenosaunee. The Peacemaker's status was also elevated to that of demigod who "was deputed by the Master of Life, who is also called the Holder of Heaven, to the earth, the better to prepare it for the residence of man, and to teach the tribes the knowledge necessary to their condition, as well as to rid the land of giants and monsters." Tarenyawagon (Schoolcraft's Hiawatha) then supposedly brought his message of peace to the "Council of the Iroquois," and thereby established the Confederacy of the (then) Five Nations. Afterwards, he and his daughter proceeded to the council, across what is now Onondaga Lake in his "magic canoe." At that point, a great eagle appeared from the sky, descended, and killed Hiawatha's daughter, simultaneously annihilating itself. Shocked and then disconsolate, Hiawatha was unsure what to do until, once more realizing the importance of his peace mission, he addressed the great council whose members were inspired by his words. With his mission completed, Hiawatha ascended to heaven in his magic canoe.[24] The appearance of Onondaga Lake is one element that seems consistent with the versions of the Hiawatha narrative derived from Indigenous sources.

Longfellow's *The Song of Hiawatha* was a huge success, perhaps the most commercially successful poem in American history. 50,000 copies were sold in the first six months. Longfellow, already a popular poet, became a famous and wealthy man. The poem was hailed as "the greatest contribution yet made to the native literature of our country."[25] Plays and pageants based upon the play's narrative were staged throughout the country. Many Americans were hungry for a story that included the "first Americans" and offered them a sympathetic, even if largely invented, role in the nation's evolving historical self-understanding. Longfellow's version was also a tragic love story that fit the romantic longings of nineteenth-century America, and which have resonated through American popular culture ever since. Longfellow turned Hiawatha's narrative—among the most complex and culturally significant for woodland Native Americans—into what would become the script for several Hollywood films.[26]

The cultural desire (perhaps propelled by a sense of guilt) to reconcile with Native Americans was strong enough that it was unshaken when contemporaneous anthropologist Horatio Hale revealed the weak evidentiary foundations upon which the poem was based. Many fewer people read Hale's work than read Longfellow's, but it represents a more reliable attempt to reconstruct the story of the Great Peace and Hiawatha's role in it. Hale's was the first attempt on the part of a non-Native American to provide a literal retelling of the story of the Great Peace. First published as an article, "A Lawgiver for the Stone Age," in the *Proceedings of the American Association for the Advancement of Science*, it was eventually turned into a monograph, *The Iroquois Book of Rites*. Hale's informants were, he stated, the "'wampum-keepers'" or "official annalists" of their people.[27] If Longfellow was fanciful and romantic, Hale was hard-nosed and scientific. He had little patience for the more symbolic or mythological elements of the Peacemaker narrative. For him, Hiawatha was a lawgiver and political leader, to be grouped with "Confucius and Solon, with Numa, Charlemagne, and Alfred, or (to come down to recent times) with the greatest and wisest among the founders of the American Republic."[28]

In the process of debunking the Schoolcraft/Longfellow version of Hiawatha, Hale carefully distinguished between three important figures in the story: first, Atotarho (Tododaho) was the chief of the Onondagas, a frightening and aggressive figure who held other woodland tribes under surveillance and threat. Next, Hiawatha was presented as an older, wise, and benevolent figure, who sought to bring peace to the Five Nations but was stymied in his efforts by Atotarho. Finally, Dekanawidah, a chief among the Mohawks, recognized the wisdom of Hiawatha's plan and helped him to execute it, first among the Mohawks and Oneidas. Together, they slowly and patiently brought the other nations together to establish the "great peace" within the umbrella of Haudenosaunee Confederacy.[29]

Hale was dismissive of anything beyond what he took to be literal historical truth. For example, he discounted references to Atotarho as having snakes in his hair, of Hiawatha ascending to heaven in a white canoe, or even of Hiawatha as the originator of wampum (which Hale contended was used by the Moundbuilders, the earliest known native people). Hale argued that the most authentic version of the Hiawatha story was told by Native Canadians, partly because their societies were subjected to less cultural trauma and fragmentation than occurred on the southern side of Lake Ontario.

With regard to Longfellow, Hale concluded that "If a Chinese traveler, during the middle ages, inquiring into the history and religion of the western nations, had confounded King Alfred with King Arthur, and both with Odin, he would not have made a more preposterous confusion of names and characters than that which has hitherto disguised the genuine personality of the great Onondaga reformer."[30] Hale's assertion notwithstanding, it is not at all clear that there is only one true telling of the Hiawatha story. His search for the definitive Hiawatha may be, in its own way, as misplaced as Schoolcraft's fictive reconstitutions. Still, in spite of all of Hale's disagreements with his poetical counterparts, what joins them is recognition of the significance of geography and place. In Hale's case, he gave a prominent place to the area's watersheds, with specific reference to Onondaga Lake. Hale determined that on its shores, the first council of the leaders of the Five Nations was convened to map out the specific rules under which the new league would be formed. In fact, from the nineteenth century onward, virtually all attempts to capture Hiawatha's historical and political importance to the Haudenosaunee, include Onondaga Lake and its surrounding environs as a central feature.

Thomas R. Henry's *Wilderness Messiah*, published in 1955, was another attempt to depict Hiawatha as a real historical figure. Henry began with the premise that "there is hardly a trace of identity between the real man and the incongruous mixture of mountain-tossing demigod and love-sick Victorian gentleman, dressed in deerskin and feathers, of Longfellow's poem."[31] Like Hale, Henry downplayed Hiawatha's magical origins and power but emphasized the horrific aspects of Ododarhoh. Henry contended that Ododarhoh killed Hiawatha's entire family, and his hair was indeed composed of snakes or serpents: "He is a master of wizardry and by his magic destroyed" men and women that came to close to him. "He ate men and women raw" and controlled the Onondaga nation by instilling fear of his magical powers. Hiawatha, having lost his family to Ododarhoh, left the community to live alone in the forest as a hermit. Here, the Peacemaker Degawandawida found him, and convinced him that it was time to bring a stop to the violence that haunted the woodlands people. Degawandawida, while a visionary, also recognized his own lack of speaking skills and turned to Hiawatha for help in communicating his message. Hiawatha thus brought Degawandawida's vision of peace to Ododarhoh, convincing him to renounce his monstrous ways. For his success in removing the snakes from Ododarhoh's hair, and by implication, smoothing the conflicts between the Haudenosaunee nations, he received his name Ayonwartha, "the comber."[32] When I interviewed Clan Mother Wendy Gonyea, she explained to me that the transformation of Degawandawida is interpreted by the Onondagas as a lesson in redemption. Even those who seem irredeemable are capable of transformative changes of the heart and of making lasting contributions to the community.[33]

The Onondagas, "amazed at the miraculous transformation, followed [Hiawatha] into the Great Peace." Ododarhoh became the "firekeeper," and the Onondaga territory became the seat of the great council fire. Hiawatha eventually convinced all Five Nations of the value of peace and for a period of time established peaceful relations with the Wyandottes, Eries, Hurons, and Cherokees. Deganawida then commanded the chiefs of the new confederacy to "uproot the tallest pine tree" and into the hole that was created "cast all weapons of war," saying, "We bury them from sight forever and plant again the tree." Deganawida then went to the shore of Onondaga Lake, where a white stone canoe waited. He boarded it and "paddled westward with his grieving followers watching from the shore as the mystic craft disappeared into the setting sun. On the other side of the lake, they say, his mortal body is buried, and his grave is covered two spans deep with hemlock boughs."[34] In Henry's reconstruction of the story, Hiawatha then traveled south into what is now Pennsylvania and north into what is now Ontario, bringing the message of peace. He returned with strings of shells which he strung into belts. These wampum belts provided the symbolic legacy of peace.[35] Then, Hiawatha paddled his canoe eastward across Lake Champlain, and was never seen again.[36]

In the twentieth century, Native American archaeologist J.N.B. Hewitt systematically collected oral histories the Haudenosaunee, in an effort to clarify the factual elements of the Hiawatha narrative, particularly in terms of distinguishing Deganawida and Hiawatha. Hewitt also uncovered the importance of Jigonsaseh, a Clan Mother, who is now generally considered a cofounder of the Confederacy. In some versions, Jigonsaseh preceded Hiawatha in recognizing the importance and value of Deganawida's message of peace.[37] Deganawida reportedly told Jigonsaseh that the Clan Mothers had a duty to replace chiefs that had wandered from the Great Law of Peace, thus reinforcing Haudenosaunee matrilineal traditions.[38]

Hewitt's investigations revealed several common elements in the multiple versions of the story. First, a period of civil war or strife among the various woodland nations was the precursor to the Peacemaker's appearance. In Hewitt's words, there was a "tangled and ferocious landscape of war" at that time. What isn't clear is whether the war resulted from internal struggles, stemming from changing agricultural practices, or from an external threat. Sometimes the conflict is characterized as an invasion from a vast empire in the south.[39] The Peacemaker then appears and is never in any version of the narrative the same person as Hiawatha. He approaches Hiawatha, a dangerous misanthrope who lives alone in the woods and practices cannibalism, converting him to the cause of peace at Jigonsaseh's behest. Hiawatha has exiled himself from human contact because he is distraught at the loss of his daughters and/or wife. The Tehyohrohnyohron, a magical bird, is in some way responsible for these deaths. For instance, in one version, Hiawatha's pregnant daughter is trampled by a crowd in pursuit of the beautiful bird.[40] Disconsolate, Hiawatha travels south where the waters of the Tully Lakes, blocking his path, are lifted by friendly ducks. On the bottom of the lake bed, Hiawatha finds sparking shells that he picks up. Gathering thrushes growing along the shoreline, he fashioned these into belts, stringing shells to create wampum. The Haudenosaunee Condolence Ceremony is grounded in Hiawatha's words, offering comfort for anyone who had suffered his same grief.[41] In some versions, the shells are found on the shores of Onondaga Lake. According to Clan Mother Wendy Gonyea, Peacemaker appeared first on the northeastern shore of the lake, to bring his message of the Great Peace.[42] The area is now designated for public use as the Long Branch Park.

After Hiawatha's acceptance of the Peacemaker's plan, the two traveled the countryside to convert the Five Nations to its message. Adodorah is transformed by Hiawatha and the Peacemaker into a human form, with the snakes combed from his hair and (in some versions) a shortened penis. From this Hiawatha earns the name "He Who Combs."[43] The Peacemaker then goes back to Quint Bay in Lake Ontario, while Hiawatha returns to live among the Mohawks.

Given the variations associated with the story of Hiawatha and Peacemaker, it is impossible to say definitively which of them may be true, or "real," a term that is sometimes used in opposition to the Schoolcraft/Longfellow version. Nevertheless, this does not force us to accept all versions as equally valid. Some are truer to Native American traditions and have a more secure grounding in archaeological findings than others. The *Encyclopedia of the Haudenosaunee*, for example, attributes validity to all Hiawatha narratives other than the Longfellow/Schoolcraft versions. Some attempts have been made to match scientifically derived evidence with narrative accounts. Astronomical records, for example, have been matched with references to a darkening of the sky in some of the narratives. Matching these with physical artifacts led researchers Barbara Mann and Jerry Fields to August 31, 1142 as the date near which many of the actual events transpired.[44] Fairly definitive evidence exists, in other words, that the Confederacy was established, not in response to European contact, but as the result of a civil war that occurred among the various nations due to internal pressures of some kind.[45] The era is designated by some Haudenosaunee chroniclers as "The Second Epoch of Time: The Great Law of Keepings." Still, some oral traditions refer to an earlier Confederacy, dating back even before the first millennium.[46]

The point of recounting these versions of the Hiawatha narrative is not to distinguish fact from fiction, the mythical and symbolic from the literal. In fact, such categories are themselves problematic. Moreover, the disentanglements involved reflect the history of Haudenosaunee subjugation as much as they do the complexities of interpreting oral narrative traditions. The main purpose here is to give an inkling of the narratives that are woven through the landscape of upstate New York, the Finger Lakes, Lake Ontario, and Onondaga Lake. The Haudenosaunee are a storied people who developed a form of governance that influenced the evolution of American political traditions (recognized by a congressional resolution in 1988).[47] The Haudenosaunee were among the first to encounter Europeans in North America. They developed complicated patterns of trade and diplomacy with the Dutch, French, and English and are among the most researched Indigenous groups in the world. The Hiawatha narrative has been central to the Haudenosaunee's political and cultural identity. Onondaga Lake, a small lake surrounded by urban and suburban development that would probably strike the casual observer as geologically or aesthetically insignificant, has had, and still has, a central place in Native American histories that stretch back nearly 1,000 years. This material has consequences that reverberate into an array of current legal and political issues.

The Arrival of Europeans, Scientific Methodologies, and Industrial Development

In the 1600s, Europeans arrived in what is now upstate New York. First were the Dutch, who established a fort in what became the City of Albany. Their interests, at least at first, were mostly in trading with the Indigenous people that they encountered. While the Dutch made forays into what they and other Europeans considered to be "the wilderness," they appear to have had little interest in conquering and colonizing large tracts of land. It was the French in the north and the English in the south who saw these lands as empty spaces whose Indigenous occupants needed to be "civilized." When Europeans began to infiltrate the lands occupied by the Haudenosaunee, they brought an entirely new worldview, shaped by Christianity and the scientific revolution. Perhaps the conflict that occurred was inevitable, perhaps not. But collision occurred, not only at the level of warfare and political subjugation, but at the level of ideas and narratives, where the modern scientific paradigm won out.

We don't know what would have happened on the North American continent had there not been contact between Europeans and Native Americans. We do know that Native Americans had established complex and harmonious societies, living in coherent, stable, and prosperous communities for thousands of years before the European invasion. The people who were here did not need the political, economic, and religious ideologies of Europeans to live meaningful lives. But once Europeans came, things changed drastically across what would become the United States. A new system of thought and management of relations between humans and the natural world took hold. Onondaga Lake's transformation from one of the most sacred sites in the continent to its sad description as the "most polluted lake in America" shows how these complicated power relationships shifted in accordance with a new ideological system. Even so, that newer system was itself unstable, and the transformations of Onondaga Lake that occurred in the twentieth and twenty-first centuries may provide some glimpse of what the future holds.

Two main versions of the origin story compete in the European context: one grounded in religion, i.e., the biblical story of Genesis, and the other grounded in scientific empiricism. While they are in conflict with one another in many respects, they overlap in others. Specifically, they are part of what Carolyn Merchant has called the "recovery narrative," that is, the belief that human history has a definitive and knowable point of origin, that its unfolding represents a form of progress, that humans have the preeminent place in the natural order of things, and that domination and control of natural forces by humans is appropriate.[48] These features distinguish the European origin stories from the Haudenosaunee cosmology, which includes none of these ideas.

Of the scientific disciplines, geology is most relevant here. For Onondaga Lake, geology was destiny, because the lake was located near a vast bed of highly prized limestone and major salt beds which provided the material resources for the economic development that contributed to the lake's decline. Moreover, the lake provided a convenient dumping ground for the toxic materials that were generated by local industries and the sewage from the expanding populations that serviced them.

Geology itself is a system of what Michel Foucault called "power/ knowledge."[49] Science is not simply the additive accumulation of facts leading toward the development of ideas or theories. Facts are shaped into theories through complex processes, as the theories upon which facts depend evolve and change over time. Scientific understanding can be subject to what Thomas Kuhn labeled "revolutions."[50] To recognize this is not to discount the value or validity of scientific findings, but even during a period when facts, especially scientific facts, are under siege, we should nevertheless avoid the trap of believing that facts are autonomous and unproblematic, because that has never been the case.

Geological investigations, while having a component of what might be called the "purely scientific," were and are involved with the enterprises of mining, quarrying, and drilling the earth's surface for profit. Geologists developed a set of techniques that could be used to locate buried fossil and mineral deposits and employ them towards industrial development. Moreover, geology's genesis as a scientific discipline occurred during the period of European colonization. It is no accident that the Solvay Process company, a central player in this story, was founded by a mining engineer.

According to Gary D. Rosenberg, the idea of landscape, which preceded the science of landscape (in essence, the science of geology), involves the notion of "continuity of space."

> The concept of spatial continuity and specifically of landscape as a continuous panorama of integrated entities—i.e., of landscape as more than the sum of its parts—is rather new in Western cultural history. Geology, the science of landscape, originated after the planet was classified alongside the other bodies revolving around the Sun and after the idea of landscape had taken shape.[51]

Geology was part and parcel of the scientific revolution, which relied heavily upon the development of geometry. Before there was landscape, there was God and the rooting of identity and knowledge in particular locales. Viewing land as a series of mathematic extensions undergirded not only the science of geology but also facilitated its division into private parcels that could be measured, defined, protected, and sold. "The word, 'landscape,'" Rosenberg explains, "became part of European languages at the same time landscape art achieved its independence, signifying existence of the concept in the mindset of the times."[52] In pre-Renaissance artistic representations, "Individual features of the landscape were viewed as isolated objects that were not incorporated into a unified whole."[53]

The founder of modern geology is often considered to be Nicolas Steno. Born in 1638 in Copenhagen, Denmark, early in his life Steno was actively involved with scientific investigations of human anatomy and geology before converting to Catholicism and becoming a priest. Steno famously concluded, on the basis of teeth he found in a head of a giant shark sent to him by a patron, that stony objects (*glossopetrae*) found in rocks that were previously thought to be stones fallen from the sky, were in fact the fossilized remains of dead sharks. Steno's work on fossilization led him to a study of geology, through which he established basic principles of stratigraphy, for instance, that inclined strata had once been horizontal, composed of a fluid that settled and solidified over lower-level strata.[54]

The first attempt to systematize geological formations into a set of categories was accomplished in the late eighteenth century by Abraham Werner. Werner noticed what appeared to be a base of crystalline rocks supporting and overlapping sedimentary ones. He concluded that this was the earth's original crust and labeled it the *Urgbirge* or Primitive series. The sedimentary rocks within it, such as limestone or shale, he labeled the *Flotzgebirge* or flat-lying series. The layer of looser fill and gravel on top of that he called the *Angeschwemmtgebirge* or Alluvium.[55] The system soon became more complex as two other layers were added to create five: the Primary, the Transition, the Secondary, the Tertiary, and the Alluvium.

Two British researchers, Roderick Murchison and Adam Sedgwick, gave names to the eras that generated these strata. Murchison coined the term Silurian for the oldest stratum that he uncovered, taking it from an ancient Welsh tribe, the Silures. Sedgwick called the oldest series that he uncovered the Cambrian, taken from the ancient name of Wales itself. Later Charles Lapworth proposed a stratum called the Ordovician, from an archaic British tribe. Another stratum was labeled the Devonian, after the town of Devonshire. Above the Devonian was the Carboniferous, the stratum where carbon fuels, coal and oil, are generally located.[56] Eventually, geologists expanded this list, dividing the earth's history into five eons, each with multiple periods, stretching back through the earth's 4.6 billion-year history.

The concepts that Murchison and Sedgwick developed were foundational and still provide crucial reference points, and thus indelibly link the scientific discipline of geology with Britain, and, by timing and implication, with the British Empire.[57] The European science of geology cannot, then, escape associations with imperialist expansion. Moreover, geology helped to provide the practical means to make imperial expansion commercially viable, while also creating a system of knowledge that competed with and eventually displaced Indigenous ways of knowing and establishing meaning. As with so many other things, in the battle for naming the ancient places occupied by the Haudenosaunee, European conquerors held the upper hand.

This is not to discount the findings of geologists. Scientific investigation and the scientific method are powerful tools for understanding the materials and forces that constitute and shape the natural world. But at the same time, the connections between scientific knowledge and imperialism cannot be ignored. The pursuit of pure scientific understanding may provide motivation at an individual level; that is, individual scientists may be primarily driven by a desire for knowledge, but the larger context in which science operates is political and economic. The knowledge derived from science is deployed for practical ends that might even be anathema to the scientists involved in the investigatory process.

From Geological Origins to Solvay Process

Having said this, the geological story of earth's development is remarkable, daunting in its complexity, and a work of investigation and imagination not always appreciated by the public for its grandeur and vision. Moreover, some of geology's most significant findings are relatively new. A major leap in understanding was made early in the twentieth century with the theory of continental drift, but a consensus about plate tectonics did not arrive until the 1950s.

In the earth's origin story as told by geologists, starting soon after its formation 4.5 billion years ago, the earth began to cool, and a hard crust developed, with crustal pieces floating across its liquid core. As the larger pieces of crust floated, they consolidated into bigger chunks or supercontinents and then split apart, in a process that continues to this day. As the crust thickened and hardened, it formed into solid cratons, which still form the foundations for the earth's seven continents. Splitting and consolidation tend to occur in patterns defined by these cratons.[58]

Evidence is sketchy for the earliest period of Earth's existence, known as the Archeon Eon, even though it lasted 2 billion years, or roughly 40% of Earth's entire history. Little Archeon crust has survived to the present time, making it difficult to draw conclusions about this era. Some geologists have speculated that as many as three supercontinents formed and divided during this period: Vaalbara (~3.6 to ~2.8 billion years ago), Ur (~3 billion years ago), and Kenorland (2.7 to 2.1 billion years ago).[59] Whether or not that is the case, what is more certain is that in the early Archeon Era, proto-continents, the precursors of today's continents, were comparatively small since the heat radiating from below prevented them from coalescing.[60] While the earliest stages of continental formation are complex and fascinating, and help to complete the original story, they are not of primary concern to the present study, which is targeted toward the geological processes that provided the foundations for soda ash production in Solvay, New York.

O. D. von Engeln states that "the geological history of the Finger Lakes began about 550 million years ago, when much of the eastern area of North America was a nearly featureless plain underlain by crystalline rocks."[61] This aligns with the start of the Cambrian. Onondaga Lake, while not itself considered a Finger Lake, lies within a large geological region known as the stable interior, which was formed during the Cambrian period. More specifically, Onondaga Lake is within the central lowland of this region, which stretches from much of what is now the American Midwest up through central western Canada. To the north is the Canadian shield.[62] To the south and east was the Appalachian orogeny (uplift) and remnants of the Grenville orogeny.

The area was covered by an ancient sea, approximately 100 feet deep, that expanded and receded over hundreds of millions of years. The Cambrian period is distinguished by the appearance of complex organisms such as corals and mollusks that evolved with protective shells with high calcium carbonate content, the elemental structure of which is calcium, oxygen, and carbon. The shells of these organisms eventually turned into deepest layers of limestone that underlie central New York State.[63]

Four hundred million years ago, in the Devonian period, the Devonian Sea covered much of New York State. Warm and relatively shallow, it teemed with life, including corals and shark-like vertebrates. The Devonian period is sometimes known as the Age of Fishes. During this period, mountains and plateaus rose in the southeastern part of the continent (then called Laurentia) and their erosions washed large amounts of shelly material, sand, and gravel into the seabed that covered what is now New York State, adding to the sedimentary materials already formed, transforming into dolostone and very hard limestone. The combination of fossilized Devonian Sea life and the eroded material from the Acadian uplift created the Onondaga formation, which stretches across New York State, running from just south of Syracuse to Buffalo, and then along the northern Lake Erie shorelines, its height diminishing as it moves westward. The resulting limestone from this period stretches as an escarpment across much of New York, and ranges from sixty-five to 200 feet thick. It is very tough and has been especially prized as a building material.[64]

A nineteenth-century geology text describes this as "the last of the great limestone formations, the Coniferous."[65] Onondaga limestone, according to Schneider, "approaches granite" in its character, and is sometimes substituted for it. "As a building stone it is unsurpassed" and has a national reputation. Moreover, "when burned this rock furnishes an extremely pure white lime."[66] St. Mary and St. Paul Cathedrals, City Hall, and the Hall of Languages at Syracuse University were all constructed of this limestone, which was quarried at Split Rock Quarry and the quarry at Jamesville.

The formation of Onondaga limestone ended when, during the Middle Devonian period, Laurentia collided with the continent of Avalon, creating a major uplift, forming the Acadian Mountains, in what is now New England. As a result, the Devonian Sea deepened in New York, and the abundance of life that occurs in shallow waters diminished drastically.[67] But the limestone that was left behind would become crucial for the siting of the Solvay Process in central New York.

During the Salina period, approximately 300 million years ago, evaporations from a vast shallow sea that covered what is now central New York created salt deposits that come close to the surface near Syracuse and provided the economic foundations for the city. These were covered by various top layers over millions of years but extended underground for long distances to the south. By the time the Solvay Process began to organize its operations, much of the salt near the city of Syracuse had been exhausted, but as one traveled farther south it could still be found underground, and was especially abundant near Tully Lake. A nineteenth-century geology text states as follows: "only three years ago, a well was sunk in the southern part of our county, and after passing through the overlaying and protecting limestones, pure rock salt was found." The overlaying rock was as much as 1600 feet deep, but beneath it was a bed rock salt sixty to 200 feet deep. Thirty wells were being drilled in Tully, and "their plan of operation is to allow the water from the Tully Lakes to flow into the wells, after it has absorbed the salt, it is forced up and sent on to the Solvay Process Company. As this brine is much stronger than that pumped up near Onondaga Lake, it is probable that it will soon supplant that article together in the manufacture of salt."[68] This process is known as "solution mining," and while it is no longer applied to the salt near Tully, it is still practiced farther south at the end of Seneca Lake, where some of the largest salt mines in the country still operate.

Above the hard limestone of the Onondaga formation lies a formation of the later Devonian period, approximately 350 million years ago, known as the Marcellus shale. Schneider notes that "we often find particles of coal, and a small percentage of combustible material, distributed throughout the mass of this shale[.]" But they are of little economic value. He notes, "We occasionally hear of fortune hunters who are wasting their time in the vain hope of finding coal in this shale."[69] Little did he know that someday these deposits would be mined for natural gas via a highly controversial and environmentally damaging process known as hydraulic fracturing.

What of Onondaga Lake itself? Onondaga, like other lakes in central New York, including the long, narrow, and deep Finger Lakes, were the result of glaciers that formed and retreated over the last million and a half years of the Pleistocene period, which ended a mere 10,000 years ago. The massive Laurentian ice sheet that invaded New York State during this period made four separate advances into the state and retreated back to Canada in between. As the ice advanced it scraped clean the overlying soils and gravels that covered the land masses, scrubbing down to the hard limestones, dolomites, sandstones, and shales underneath. Onondaga Lake lies in a limestone valley, known as the Onondaga Trough, that was scraped clean by the ice sheet and filled with glacial water as the ice retreated.[70] Its southernmost tip is defined by the Valley Heads Moraine, near the village of Tully. The Tully Valley defines the southern end of the watershed and contains Onondaga Creek, which runs about twenty miles into and through the city of Syracuse, joining its west branch about seven miles north, until it becomes a feeder for

Onondaga Lake.[71] The ice sheets reformed the entire watershed of New York State, sculpting large drainage basins. Onondaga Lake lies in the Ontario–St. Lawrence Basin, and like the Finger Lakes, its waters enter from the south and drain northward eventually into Lake Ontario. New York is a state blessed with rivers and lakes, and the complex and intertwined system of lakes, rivers, and streams along the northern part of the drainage area were highly regarded by the area's original settlers for their sacred and practical value.

The geological formations that created Onondaga Lake are not simply a matter of abstract scientific interest. The underlying structures of the area helped to provide the material for economic development and growth. The Erie Canal, sometimes referred to as the "the ditch that salt dug" was dredged along a path to facilitate the transport of salt from Syracuse to points east and west for use in food processing (e.g., salt pork) and industrial production.[72] It was instrumental, as I will show in some detail, in locating the Solvay Process on the shores of Onondaga Lake. Geological features also determine the flow of water, and hence the flow of pollution, into and out of the area. Water moves slowly through Onondaga Lake. What hydrologists refer to as the "flush rate" is relatively low. As a result, once contaminants are concentrated within it, even if the contamination stops, they tend to remain there, especially if they have heavy atomic weights, including heavy metals like mercury and chromium. Moreover, the complexity of the region's watersheds, its numerous creeks and streams, many of which are referred to in the Great Peace narrative, provided ample opportunities to dump wastes, which would in turn provide multiple entry points for contamination into Onondaga. The Seneca River, which provides drainage for Onondaga Lake and the Finger Lakes, offered a tempting solution to these problems, given its higher volume and flush rate, a solution rooted in the geological formations of the region, but one that largely ignored the region's ecological and cultural integrity. Its deployment for such purposes, however, turned out to be impractical.

Today Hiawatha's legacy is embedded in the public and commercial landscapes of Onondaga County: Hiawatha Point Park, Hiawatha Boulevard, Hiawatha Heights Apartments, Hiawatha Used Cars and Auto Parts. And thanks in large part to Henry Wadsworth Longfellow, Hiawatha's name also appears in the upper Midwest. In Wisconsin, you can find Hiawatha Hobbies, the Hiawatha Bar and Grill, the Hiawatha Resort, and the Hiawatha Golf Club. Perhaps most striking and most visible is the Hiawatha National Forest, a vast area of almost one million acres in Michigan's Upper Peninsula. While the forest is a somewhat protected area and includes such striking geological features as the Picture Rocks National Shoreline, it is also a testament to the fractured reconstructions of Indigenous narratives that European Americans have casually undertaken, sometimes, as in the case of Longfellow, for considerable financial gain.

2

Transforming a Landscape

The Chemical Revolution and the Solvay Process

The upstate lands, lakes, and rivers that European colonists took by force for settlements, starting in the seventeenth century and continuing well into the nineteenth, would eventually provide the material foundations for the American Industrial Revolution. Thanks to ample waterways and transportation, including the completion of the Erie Canal, upstate New York cities expanded significantly as new industrial forces developed around manufacturing and production. In the middle to latter parts of the nineteenth century, a chemical revolution brought forward the development of various useful, but often toxic, substances at a remarkable rate.

The first stage of the Industrial Revolution involved the expansion of mechanized technologies such as spinning looms and continued with the application of steam to power machinery. Coal, of course, had a central place in this process. But if mechanization was central to early industrial development, increasingly, as the nineteenth century proceeded, chemistry was critical for its continuing progress. Along with the transference of energy through machines, solvents, acids, alkalis, and other chemical agents were necessary to turn raw products into finished goods. One such crucial chemical substance was soda ash. The Solvay Process for the production of soda ash, although little known outside of chemical engineering circles, was central to nineteenth-century industrialization. Soda ash is more formally known as sodium carbonate, and its uses in Europe can be traced back to ancient Mediterranean cultures, including the Egyptians and Greeks, who used it as a cleaner and for various medicinal purposes. Later it was employed in dyeing, glass making, and soap production.[1] Sodium bicarbonate, a less caustic chemical commonly known as baking soda, also has multiple applications both in the home and for processes of industrial production. In small amounts it can be used for cooking, teeth brushing, or treating acid indigestion. It is also used as a water softener and as an additive to swimming pools to balance the acidic PH generated by chlorine. In glass manufacturing, soda ash is used as a "fluxing agent," that is, it allows the silica, a core substance in the process, to melt at a lower temperature, thereby significantly reducing the amount of energy needed for production. About half of soda ash is currently used in glass manufacturing.

While soda ash occurs in a form called trona, which can be inexpensively mined and processed, sources of trona are limited to a few parts of the world. No significant trona deposits exist in Europe, for example, the original home to the Industrial Revolution. While soda ash was abundant in Egypt, transporting it from those mined deposits was not economically feasible. Soda ash can also be extracted from burned trees, but while trees have high levels of potassium, which yields a useful substance called potash (used as a fertilizer), only modest amounts of soda can be extracted from them. As demand for soda ash increased in the seventeenth and eighteenth centuries, the amount of burned wood necessary to yield soda ash was so great that other sources became necessary.

The ashes of some sea flora do yield high amounts of sodium. Especially useful is barilla, a variety of kelp, mostly indigenous to the Mediterranean, particularly along the coast of Spain, but also farther north off the Scottish coast. Scottish barilla is exceptionally high in sodium carbonate. As a result, in the eighteenth century, Scottish kelp became a major source of soda, critical for the advancement of French glassmaking. Kelp was valuable enough that Scottish farmers took time away from their fields to gather it from along the shore. However, as antagonism increased between the French and British in the late eighteenth century, the French needed new sources that would be less reliant upon British imports.[2]

The French situation was considered dire enough that King Louis XVI offered a generous cash prize to anyone that could develop a process for extracting sodium ash from salt. Sodium and chlorine are tightly bound in salt molecules, however, making their separation extremely challenging. The first person to develop a practical process was French physician Nicolas Leblanc. The Leblanc process, as it came to be called, involved many stages of production and yielded a high proportion of toxic byproducts. First sulfuric acid was mixed with the salt and then heated in cast iron furnaces to create sodium sulphate, known as "salt cake" or Glauber's salt (named for the Dutch chemist Johann Glauber, who discovered it in a natural state in spring water). Sodium sulfate has industrial applications by itself and can be used in glass making to help with the removal of bubbles. In the Leblanc cycle, sodium sulfate is heated with limestone to create "black ash." The black ash is then mixed with water. Carbon dioxide is blown through this liquid, precipitating out any calcium and carrying away the sulfur (as a gaseous form of sulfuric acid). Zinc hydroxide is added to precipitate any remaining sulfides. The liquid is separated from any precipitates and dried via a heating process. The remaining ash is again dissolved in hot water, with any residual substances that did not dissolve being removed. This solution, when cooled, crystallizes pure, or nearly pure, sodium carbonate. Sodium hydroxide, aka caustic soda, which also has a wide array of industrial uses, such as for bleaching in the papermaking process, can be derived by treating soda ash with lime in a process known as causticization.

The Leblanc process, while laborious and complex, yielded a number of chemical substances that were extremely useful for industrial development. Its disadvantages included the high temperatures needed in the process (making it expensive), and the necessity of storing high volumes of sulfuric acid, a substance which off-gases hydrochloric acid and carbon monoxide, both of which can be lethal to humans and other living beings.[3] Large amounts of solid waste were generated, including hydrogen chloride and calcium sulfide, much of it spread near fields where the Leblanc plants operated. Residents near such facilities objected, spurred partly by the noxious smells that resulted from calcium sulfide waste, which as it reacted with the atmosphere produced hydrogen sulfide, which smells like rotten eggs.[4]

Dr. Leblanc did not benefit from his discovery. While the royal government offered him a fifteen-year patent, after the overthrow of Louis, the new regime seized his plants and made his patent, along with others, public property. By the time Napoleon Bonaparte restored his claim, Leblanc was destitute and had no capital to restart production, and at this point, there was no putting the genie back on the bottle. Protected from lawsuits for patent violations, industrialists in England and throughout Europe had begun setting up Leblanc manufacturing facilities.[5] Broke and distraught, Leblanc committed suicide in 1806.[6]

In an ironic turn of events, England, by embracing the Leblanc process, became the world leader in synthetic soda ash production as its engineers continued to refine it. England had an abundance of the raw materials used in the process—coal, salt, and limestone—and the government left their extraction and use untaxed. But the toxic byproducts that resulted from the process were difficult to control or dispose of, including nitrogen oxides, hydrochloric acid, manganese, sulfur compounds, chlorine, and soot from burning coal. The Tyneside, Merseyside, and Clydeside districts near Liverpool, from which the finished product could be easily shipped throughout Europe, became notorious for their heavy black smoke.[7] Those unfortunate enough to live in proximity to a facility engaged in the Leblanc process were drastically and intensely affected by it, to the point that Leblanc sites are still being studied for the continuing negative environmental effects that they produced.[8]

In response to public outcry, the British Parliament passed the Alkali Works Act in 1863. The act was specifically designed to deal with the hydrochloric acid problem. It did not prevent emissions per se, but appointed an inspector to encourage abatement, and also provided funds for researching productive uses for the harmful byproducts. It was one of the first environmental laws passed in Britain, and it still exists in a revised form. A German chemist, Ludwig Mond, who eventually took British citizenship, developed the means to extract pure sulfur from Leblanc byproducts, thus mitigating some its environmentally damaging impacts, and thereby insuring the continued economic viability of the process. Mond's efforts did not fully pay off, however, until the market for chlorine, another Leblanc process byproduct, dramatically expanded in the 1880s, as it became widely used as a bleaching agent.[9] Mond would eventually join forces with Ernest Solvay.

Even as the Leblanc process expanded and was refined, chemists were convinced that another, cheaper, less complex, and less toxic method for deriving soda ash was possible, which involved using ammonia in combination with brine. Augustin Fresnel, a French physician, is remembered primarily for his contributions to physics. Of particular note were his studies of light refraction and polarization.

Fresnel lights are still used in theatrical production.[10] But he also showed that under laboratory conditions, sodium bicarbonate could be precipitated from a solution containing brine and bicarbonate of ammonia. Several failed attempts were made to apply Fresnel's theory to practical industrial applications. From the late 1830s through the 1840s, at least ten chemists attempted to build profitable ammonia process devices, with all failing for various reasons, including their inability to minimize waste products or to produce enough soda ash to be profitable. One problem with the use of ammonia was that once started, it could not be halted without a major disruption to the manufacturing process. The multistage Leblanc process, on the other hand, could be interrupted at any point without causing significant disturbance. Any ammonia process would have to support continuous operation, which increased the technical difficulties associated with improving it. Theophile Schloesing and Eugene Rolland came closest, when they established a factory near Paris that produced twenty-three tons of soda ash a month, a significant amount. However, their salt requirements were much greater than those required for the Leblanc process, the salt waste left over held no useful purpose and was taxed by the French authorities, which further undermined the operation's profitability.[11]

Ernest Solvay was, in some respects, an unlikely candidate for developing a major breakthrough in industrial chemistry. His family lived in Rebecq-Rognon, a small village in the Walloon region of Belgium. His father taught in a boarding school until he took a position to run a quarry in the area, which seems to have been crucial in piquing his son's interest in minerals and chemistry, an interest that was reinforced by Brother Macardus, one of the teachers in his Christian boarding school. But health problems prevented the young man from pursuing a degree in engineering after graduating high school, so he continued to teach himself in the sciences. Without a degree to indicate a formal scientific education, young Solvay was forced to take a job as an accountant, which he found stifling. When his uncle offered him a position in his coal gas company, he took the offer and became manager of the ammonia division, where he made important innovations, such as developing a method to concentrate ammonia in water so as to facilitate transportation.

What was then known as "manufactured gas" or "coal gas" was produced by heating coal. It was used for fuel and lighting and was in high demand even through the 1950s, when natural gas production replaced it.[12] One of its byproducts was ammonia, for which there were limited markets. Solvay's position at his uncle's firm provided him with opportunities to experiment with ammonia, and it provided a rationale for him to continue his studies in chemistry.

While experimenting with sesquicarbonate of ammonia, a chemical long used as a raising agent by bakers, he noticed that soda precipitated from a solution of brine, ammonia, and carbonic acid. He immediately understood its implication and filed a patent for his new "discovery," although this was in fact at its base the same constituent process that had excited industrial chemists in preceding decades, each time leading to a technical and economic dead end. Unaware of these previous attempts, and perhaps thus unencumbered by doubts, Ernest Solvay believed himself to be on the verge of making an important, and potentially profitable, contribution to the field of industrial chemistry, an insight that turned out in the end to be correct.[13] According to Solvay's biographer, he did a quick calculation, which determined that he could potentially reduce the price of soda from 700 francs a ton under the Leblanc process, to as low as 150 francs a ton using an ammonia method.[14]

Partly because Solvay believed that his new discovery might be relatively free from obstacles as it was put into actual production, his faith in the process helped him draw others into the enterprise. The Solvay family became a strong support network. His brother gave up his position chartering commercial freight to join the company, contributing technical, administrative and management skills. He also secured the participation of his schoolmate Louis-Philippe Acheroy, who performed experiments to improve the process. Solvay's father Alexandre provided the initial seed money. When that was exhausted, he was able to secure the services of Eudore Pirmez, a corporate lawyer and financial benefactor, who provided invaluable legal services. When Pirmez discovered the existence of several patents as part of previous attempts to use an ammonia process to yield soda ash, he advised Solvay to patent the specific systems he used in detail, rather than the general chemical processes. The result was an unassailable patent that provided Solvay the legal protection necessary to continue.[15] Pirmez also helped raise capital, putting in his own money and securing funds from members of his family. These two accomplishments did not mitigate the discouraging discovery that others had tried and failed to turn the chemical principles into a viable and profitable business enterprise, however.

Having the technical knowledge and levels of economic support that had been unavailable to previous entrepreneurs did not guarantee Solvay's success. The partners built a facility at Couillet, near Charleroi in the heart of Belgium's glass manufacturing region, and because it faced numerous technical problems, it was unable to generate production at a level that was competitive with Leblanc facilities. One of the key problems was that the machinery used to mix the chemical components would invariably become clogged. Various attempts to address this issue were unsuccessful. In fact, Solvay and his partner struggled for four years to perfect the process, facing the possibility of bankruptcy on more than one occasion. Family members worried that he might attempt suicide.[16]

The innovation that saved the Solvay company was the introduction of vertical towers to facilitate the exchange of materials. Three eighty-foot towers were utilized in the most advanced version of a Solvay plant. The process is relatively complicated, because ammonium bicarbonate must first mix with brine (water and sodium chloride) to yield the sodium bicarbonate. In the beginning stage of the process, purified brine is pumped into an "ammoniating tower," where dripping brine is mixed with heated ammonia gas. The ammoniated brine is cooled and sent to a second tower where it is mixed with carbon dioxide that has been derived from roasting limestone (creating quicklime). This mixture is then sent to a third tower where it is again treated with carbon dioxide, resulting in the precipitation of sodium bicarbonate, with calcium chloride as a byproduct. The ammonia is reusable, and is returned to the beginning of the process. The sodium bicarbonate can be heated to yield sodium carbonate and carbon dioxide, the latter of which can again be used in the process. The ammonia recovery was important to the process's economic viability, since it removes the need to continuously purchase large amounts of ammonia in order to produce soda ash. Ammonia recovery was the key breakthrough to soda ash production contributed by Ernest Solvay and his associates.

The Solvay process faced other obstacles in its competition with Leblanc operations. Leblanc production could be small-scale and flexible, and facilities could be set up close to or even as part of glassmaking operations. Moreover, Leblanc production yielded byproducts that had other industrial uses, including caustic soda (lye, useful for papermaking), sodium bicarbonate, and chlorine. Solvay operations had to be large to be profitable, because they gained their advantage from scale and cheaper costs. What allowed Solvay to make inroads into an established market was a continuing drop in his pricing and higher levels of purity.[17]

As Solvay refined the process, he made inroads into other European countries. From France, he expanded to England, then Germany and even Russia, making the company one of the earliest European multinational corporations. Solvay's expansion was possible with the help of willing local partners, such as Ludwig Mond, the German chemist living in England, who had patented a means to extract sulfur from Leblanc process byproducts.[18] As with Leblanc operations, there was resistance at some potential sites, due to the high levels of waste generated by the process. One source casually noted, "It is uneconomic to recover the calcium chloride from the sludge leaving the bottom of the tower. This waste is harmless and is usually dumped in rivers or in the sea, whichever is more convenient."[19] Area residents weren't always as sanguine. Proposals for a plant in the Alsace-Lorraine region of Germany were delayed for ten years due to local concerns about waste generation.[20] Significant opposition also arose in the small city of Bernburg in the Anhalt region of Germany. The city's mayor led the charge against a potential plant due to fears of the pollution that would be generated and other disruptions in their peaceful community. Downstream residents on the Saale and Elbe rivers also joined in the resistance. The Anhalt Parliament agreed to sell the land for siting the facility by only a single vote.[21] In spite of opposition in some quarters, in addition to other setbacks, the Solvay Company increased its production of sodium carbonate between 1863 and 1892 from 150,000 to 1,760,000 tons, as the price fell dramatically. Solvay was making the Leblanc process obsolete.

William B. Cogswell brought the Solvay process to the United States. Cogswell was raised primarily in Syracuse in the 1850s and attended primary and secondary schools in the area. While at school, he also took

Figure 2.1. The Solvay Production Facility (c. 1905). William Cogswell and the Creation of the Solvay Company. Courtesy of the Library of Congress photo archive.

lessons from Syracuse architect Luther Gifford, and was skilled enough by the time he was twelve that one of his drawings was used as the basis for constructing the Globe Hotel in Syracuse. He attended Rensselaer Polytechnic Institute for three years, studying civil engineering, but left before completing his degree. Instead, he became an apprentice at a machine shop in Lawrence, Massachusetts, a manager for machinery at Marietta and Cincinnati Railroad in Ohio, and then manager of a foundry in St. Louis. He returned to Syracuse in 1860, helping to start the Whitman and Barnes Manufacturing Company, which initially specialized in farm tools and mower blades. Cogswell worked there until the Civil War intervened.

During the War, Cogswell developed an innovative system, essentially a floating machine shop, for repairing naval ships while they were at sea, allowing them to continue participating in blockades, even while undergoing repairs. Eventually he was transferred to the Brooklyn Naval Yard, where he became manager for all machine repairs. After the war, he supervised construction of a blast furnace for the Franklin Iron Works and helped to complete the Clifton Suspension Bridge at Niagara Falls. Having been offered the position of overseeing the Rowland Hazard lead mine in Missouri, he spent five years at Mine La Motte.[22]

Cogswell learned about the ammonia soda ash process at a mining conference in Philadelphia in 1878. When a Pennsylvania geologist named Oswald J. Heinrich read a paper on the subject, Cogswell's interest was piqued. As a native of Syracuse, he was well aware that given its abundance of salt and limestone, raw materials for the process, the area was ideally suited for establishing a Solvay facility. Syracuse was known as the "salt city," because of the salt industry that had provided the economic basis for its early growth. Syracuse also had large deposits of limestone and was blessed with lakes and rivers that would provide a convenient dumping ground for calcium chloride wastes.

Cogswell traveled to Hanover, Germany, where he met with a German industrialist who was also interested in exploring the possibilities for production, but who reportedly "went insane" before negotiations on a possible partnership could be completed.[23] However, Cogswell learned of Ernest Solvay's successful attempts to manufacture soda ash via an ammonia process, and traveled to Brussels to meet with him.[24]

Ernest Solvay did not immediately embrace the chance to open an American franchise for his process. At first, he ignored Cogswell. Undeterred, Cogswell sent letters of recommendation from Cornell University President Andrew White and the prominent Syracuse lawyer and politician Charles Sedgwick. Businessman Rowland Hazard II, who eventually became a partner in Cogswell's new enterprise, also appealed to Solvay. Convinced by the many statements of support, Solvay agreed to work with Cogswell, but he drove a hard bargain. He contributed only a third of the capital, which amounted to $300,000 total, but demanded (and received) one half of any profits generated beyond 10% of his contribution (in other words, half of all profits above $10,000).[25] In return, Solvay agreed to provide plans for the facility, "to interchange information and exchange visits, allowing the men who had signed contracts to study at their plants."[26]

The first site considered by Cogswell was in Jamesville, New York, a small village about five miles southeast of Syracuse, where large deposits of limestone were known to be accessible. Drilling explorations, however, found no salt, so this site was abandoned (although it would eventually be the site of a quarry that would supply limestone to the Solvay plant).

Cogswell also considered siting the facility in Aurora, New York, on Lake Cayuga, a pristine Finger Lake, but when no accessible salt was found, it too was abandoned. The company eventually decided upon a 7.85-acre site, about a half mile west of Onondaga Lake, close to a railroad line, and abutting the Erie Canal, making for accessible transport. The site was also close to the Syracuse Salt Works, which could provide salt until the company could supply its own. According to Solvay engineer Edward Trump's unpublished early history of the Solvay works, "In the center of [the property] was a depression about 15 feet deep, with a large spring at the bottom, surrounded with some fine trees and a small house—the home of a widow Mrs. Eliza Throop."[27] While the land was undeveloped, it was also close to Blast Furnace Road, so named because of the pig iron smelter located that, and the future site of Crucible Steel. In other words, the area, while still mostly undeveloped, was in the process of become a site for industrial production. Robert, William, and Charles Gere owned and farmed much of the land on the western side of Onondaga Lake. Robert Gere was also involved with the salt business and local manufacturing concerns. He agreed to a price for the parcel with Cogswell but raised it substantially the night before the exchange was to take place. Cogswell paid the higher price, but he banned any of the Gores from ever holding stock in Solvay Process.[28]

The Solvay Process company was incorporated on October 12, 1881 and began operation in 1884. Cogswell was well-liked by the men that he hired to build this highly complicated facility, at least according to the engineer Trump. "Mr. Cogswell," he wrote, "had a faculty for picking out men for places he had to fill, who were capable and efficient. He believed in paying them well, and was always devising plans like the Participation Contracts [a form of profit sharing] to make them interested in their work, and to foster co-operation between them." The result was an "Esprit de Corps," which helped greatly during "the first hard years of development of the company."[29] One of the most important additions to the company was engineer John E. Street, reputed to be a mechanical genius, who was given the task of turning Solvay & Company's plans into an actual operating soda ash manufacturing facility. The production workers were mostly "old salt boilers, used to working on the salt kettles," who had been retrained for working with the Process.[30] These were hardened men, accustomed to working with hot, corrosive substances. Nevertheless, while salt work might have provided some preparation for process work, the dangers and difficulties imposed by the latter may have surprised even them.

The first years of developing the company were indeed difficult. Technical snags impeded production. Some of the iron rings used in the large towers, which were shipped from Belgium, were defective, and had to be replaced by American-made ones. The ammonia supply was also a problem. Recycling diminished the amount needed but did not eliminate it entirely. In European Solvay plants, ammonia was drawn from coke ovens, but none of these existed in the U.S., so a means had to be found to recycle it more effectively within the process itself.[31] The brickwork for the building was shoddily done, and the layout of the buildings turned out to be a mistake, since it caused crowding as the size of the operation grew. Valves were a major problem, because they became quickly corroded by the ammonia and lime that were being pumped through the piping system. The pumps kept breaking down, so that the engineering crew often had to be awakened in the middle of the night to try to patch them back together. The engineers had to work on pumps that were filled with corrosive substances. As Trump notes, "To put [the] cap screws back on with R. H. liquor from a leaky discharge cock over one's hands was not a pleasant or quick job."[32] When the finished product, soda ash, was finally dumped onto a basement floor, it was extremely hot and once it cooled, had to be shoveled by hand into wheelbarrows that were pushed into an elevator to take them to a first-floor packing room. The conditions in this part of the plant were positively Dickensian. As Trump put it, "The heat and dust made this cellar an inferno, and it was very difficult to keep men working there."[33]

At one point, while testing for the presence of bicarbonate crystals in one of the towers, several men noticed a strong smell of ammonia. As they investigated, they were overcome with carbon dioxide gas, which knocked two of them out, one of whom was Cogswell. Men rushed in and out of the facility, rescuing others who were being exposed to the gas. Some showed true heroism by putting the safety of their colleagues before their own. The first man received a head injury and Cogswell a black eye before they were both revived by opening a door to let air in. One worker, who had been up above the tower putting salt into the brine, was found unconscious, and the others were unable to revive him. Trump notes, "The force of men which we had trained had been severely tested by this incident, and the bravery exhibited by the Foreman and others, in repeatedly returning to the fatal tower in their efforts to rescue those overcome, showed we could depend upon them."[34]

Once the plant started operating, it became clear that the cooling system, which involved pouring water over the outside of the tower, was inadequate, because it yielded "mushy" bicarbonate that couldn't be dried to create sodium carbonate. To mitigate this problem, Cogswell and the other engineers developed a system of cooling pipes that were wrapped around the inside of the column.[35] Over time, these issues were sorted out. Trump documents a slow but steady rise in production, but another explosion occurred, which was so powerful that it blew the roof of one of the plant's buildings, injuring some of the workers and killing a "tramp, who had secreted himself in the top of the building."[36] It took two years of hard work under dangerous and difficult conditions before the plant was able to turn a profit. But, once perfected, the ammonia soda process had a major impact on global soda production. In 1863, 150,000 tons were generated globally using the Leblanc process, but by 1902, that figure had jumped to 1,760,000s, with virtually all of it created by the Solvay process.[37]

Cogswell's combination of mining and mechanical skills was invaluable for making the siting decision and creating the infrastructure necessary to run the plant. He correctly predicted that a large bed of salt lay near the Syracuse salt springs that had been the basis of Syracuse's salt industry, and which had given Onondaga Lake its unique ecological features. In 1888, a huge vein of rock salt was discovered near the village of Tully, which lies approximately twenty miles south of Syracuse. As the Solvay Process expanded, and ever larger amounts of salt were needed, Cogswell developed a system of pipes that pulled water out of two-mile-long Tully Lake and pumped it into dozens of wells drilled into the salt beds, where it dissolved the salt and created brine. The brine was then sent it to the Solvay facility to feed the soda ash process. (This method of salt mining left large empty caverns into which the hillside surrounding the valley began to subside, generating a host of environmental problems to be discussed in subsequent chapters.)

Figure 2.2. Solvay Process Lime Towers (1986). Courtesy of the Library of Congress photo archive.

Cogswell also set up a cable basket system to bring limestone from the Split Rock quarry, several miles southwest of the facility. The quarry was used up by 1912, and replaced by the much larger Jamesville quarry, which was farther from Solvay, making the system impractical.[38]

The Semet-Solvay Company and the Expansion of Solvay Operations

In 1895, the Semet-Solvay Company was formed as a spin-off of the Solvay Process.[39] Its purpose was to set up a coking operation that would involve a significant expansion of the original facility. Coke is coal that has been heated, or "cooked," to drive off the impurities that it contains, leaving a purer form of carbon that burns hotter and cleaner than unprocessed coal. While earliest forms of coke production are traceable to China, it was first used in England by brewers seeking a heating method that would not produce the smoke generated by wood or the sulfur from the burning of coal. Clean burning coke provided an alternative and resulted in what became known as "pale ale." From early on, coke producers recognized that its byproducts might have useful and economically valuable purposes but were unable to make use of these while maintaining the integrity of the coking process until the late nineteenth century. As John Fulton notes in *Coke: A Treatise on the Manufacture of Coke and the Saving of By-Products*, "The efforts for supplementing the profits of coke making, by saving the by-products of tar and ammonia from the gases discharged from the coke ovens, occupied the early attention of coke manufacturers." First efforts to capture these elements, in a facility in Sulzbach, Germany in 1766, were "very crude and of little practical value." The problem lay not only in the primitive and inefficient nature of the process, but also in the fact that there were no markets for the ammonia—a principal by-product. As a result, "Very slow progress was made in the saving of by-products until about the year 1883."[40]

The Knab oven, constructed in France in 1856, made use of the coking off-gases to run the ovens themselves, and in 1862, its operators, Simon and Carves, improved upon the oven by finding a means to scrub the by-product gases, removing the tar and ammonia, and allowing a purer heating gas to run the ovens. But it was German Albert Heusner who made the process profitable, building a hundred ovens in Germany and effectively establishing the coke byproducts industry.[41]

With the rise of the Solvay process, the market for ammonia expanded dramatically. Solvay facilities could not only use the hydrocarbon off-gasses of coke to run their coke ovens, but could use the ammonia in the Solvay process itself, and find markets for tar byproducts as well. As a result, the Semet-Solvay oven was developed, which Fulton describes as a "plain, economical oven, well-adapted to the saving of by-products," which was its main purpose.[42]

Thomas Morris, an American engineer, traveled to Belgium to learn how to construct such a facility and brought twelve ovens into production in 1896. Louis Semet, Ernest Solvay's cousin, had been responsible for developing the process, and Semet-Solvay set up coke ovens across the U.S. Other byproducts produced by the process included coal gas, which was used to fire the coke ovens themselves, and tar. The production of tar led to the forming of the Allied Chemical company from Solvay Process and four other companies, one of which was the Barrett Company, a roofing company that also was an early developer of highway tar.[43] While much of the material from the process had economic value, it also generated wastes that ended up in proximity to the production facilities.[44] Coke production generates toxic off-gasses that can damage the health of workers. Moreover, the structures were lined with asbestos, which can ravage lungs exposed to it over time with mesothelioma, a disease which kills its sufferers in an average of 18–24 months.[45] Health problems associated with working in coking facilities are now well-documented.[46]

The Solvay process arrived at the start of America's Gilded Age, a period of unparalleled industrial expansion that spawned a nearly endless demand for soda ash. By 1897, Solvay Process had become the largest soda ash producer in the world.[47] Rowland Hazard served as president of the company, with Cogswell as general manager. The production facility generated large amounts of calcium chloride waste, but there seemed to be plenty of vacant land in the area, and plenty of room in Onondaga Lake. No one noticed, or sounded any alarms, about the accumulation of wastes, at least at first. No local opposition appears to have been expressed. Had calcium chloride been the only waste product dumped into Onondaga and its watershed, the problem might never have reached the crisis point that it eventually did. But other factors would eventually intervene to intensify the ensuing environmental problems. The first involved the introduction of the Castner process to the Solvay facility. The second involved Onondaga County's need for sewage disposal.

Dynamos, Castner-Kellner, and Mercury Contamination

As noted earlier, one of the limitations of the Solvay process in comparison with the Leblanc's was that the latter yielded caustic soda and chlorine as byproducts which had economic value. Caustic soda, or sodium hydroxide, is used in the papermaking pulping process, in which wood fibers are separated from one another. Chlorine, a deadly poisonous gas, has uses as a bleaching agent for paper and textiles. The 1880s saw breakthroughs in industrial electrolysis, that is, sending an electrical current through brine. Electric dynamos increased their power outputs, which drastically reduced the production costs of caustic soda and chlorine. While caustic soda was in high demand, chlorine was less so, initially. Moreover, it was a dangerous byproduct that could not simply be released into the atmosphere. Two separate but related developments occurred at the end of the nineteenth century to make salt electrolysis more attractive to investors, especially to those running Solvay process plants, who already had access to large amounts of salt. The first was a steady expansion of chlorine markets. The second was the development of the Castner-Kellner process. While electrolysis could separate sodium and chlorine in laboratory conditions, once separated, the substances were too volatile to be of practical use. Castner and Kellner solved this problem via the introduction of mercury as a stabilizing agent.

The larger contexts of scientific discovery during this time laid the groundwork for Castner-Kellner. Sir Humphry Davy, a scientific polymath, was one of the first scientists to engage extensively with electrolysis, which he applied to numerous compounds, in the process discovering a number of highly reactive metals, including potassium, sodium, calcium, magnesium, boron, and barium. Davy often worked with Michael Faraday, whom he hired as an assistant after an accident in his laboratory damaged Davy's eyesight. Davy is perhaps best known for his invention of the Davy Lamp, a safety lamp that became widely used by English and other coal miners.[48] Faraday, one of England's most prolific scientists, continued work on electrolysis on his own, and eventually developed the electric generator and the electric motor, providing the foundation for widespread practical applications of electrical generation. Albert Einstein considered him a forerunner, given his work on magnetism and light.[49]

James Watt, inventor of the steam engine, and another of the nineteenth century's great industrial innovators, continued work in this field and found that he could isolate sodium from chlorine by running an electrical current through brine. He could not, however, develop the process for practical use, because he could not keep the highly unstable chlorine from reuniting with the sodium or impurities in the brine. Moreover, even if he had found a way to stabilize the process, the levels of electricity generation needed to make electrolysis work at an industrial scale were simply unavailable. In spite of this, he did patent his process in 1851.[50]

By the late nineteenth century, breakthroughs by German engineer Werner von Siemens and Belgian engineer Zénobe Gramme, significantly improved the powering capacity of dynamos, making an economically viable, large-scale electrical generation possible. Siemen's company was the first to build an AC generator powered by water, used to run streetlights (in the English city of Godalming, in 1881).[51] Gramme was the inventor of the Gramme machine, which was the first to produce very high voltages in direct current.[52]

Two engineers, Hamilton Castner, an American, and Karl Kellner, an Austrian, originated the mercury cell electrolysis process. As David M. Kiefer notes, "A convincing case can be made that the U.S. chemical industry came of age during the final decade of the 19th century. For the first time in history, American scientists played a key role in the development of new chemical technology." One of these scientists was Hamilton Castner. Castner, with a degree in chemical engineering from Columbia University and working on his own, hoped to start a business generating sodium cyanide, which is used for electroplating and purifying gold ore. To do this, he needed high quality pure caustic soda, which led him to experiments with electrolysis and eventually to the development of the mercury cell technology. Kellner, working separately, also developed a form of the technology. In order to avoid a long and costly set of lawsuits, they joined forces to form the Castner and Kellner Alkali Company in 1885.[53]

With the improvement of dynamos, and given greater demand for chlorine as well as sodium hydroxide, the Caster-Kellner mercury cell electrolysis system appeared at a propitious moment. It works as follows: Three chambers sit within a rectangular iron enclosure, at the bottom of which is a pool of mercury. Brine is introduced into the outer chambers, into which is also placed a graphite anode. A current is discharged, flowing from the anode to an iron cathode in the central cell, and through the mercury, which also acts as a cathode. The electrical charge separates the sodium and chlorine in the brine, with the sodium dissolving into the mercury to create what is called a "mercury amalgam," with chlorine released as a byproduct. A gentle rocking of the entire mechanism moves the mercury amalgam to a central chamber filled with water, and as it does so, the sodium in the amalgam bonds with the hydrogen in the water to create a sodium hydroxide solution, which is then removed from the chamber and replaced with more water.[54]

From an environmental perspective, the main problem with the Castner-Kellner process lies in its use of large amounts of mercury. While the mercury was recycled, over time it became contaminated with enough impurities that it could no longer act effectively as catalyzing agent, at which point it was unceremoniously released into the local environment. When the process was eventually introduced in the Solvay plant, the mercury was released into streams that fed into Onondaga Lake. Ernest Solvay had originally showed little interest in sodium hydroxide as a sideline, but once introduced to the Castner-Kellner process, he saw its utility, not only for generating a marketable product, but for helping to drive Leblanc operators out of business. Via a complex but ultimately successful series of negotiations with Castner and Kellner, Solvay eventually obtained access to the process, which rang the death knell for Leblanc producers.

Even though the Solvay process was less toxic than Leblanc's, it still generated large amounts of contaminants. In fact, it is estimated that six million pounds of salty wastes, chlorine, sodium, and calcium, were discharged into the lake from the 1880s until the plant closed in the 1986. It altered the very landscape of the lakeshore due to the salty deposits that it generated. The Solvay wastebeds, which are located near the mouth of Nine Mile Creek, remain a continuous source of carbonated salts entering into the lake and making their way into the Seneca River, where they track their way along the path said to have been taken by Hiawatha in his canoe as he paddled from Cross Lake to Onondaga Lake. Moreover, as we shall see, the chemical composition of the waste allowed it to react with phosphorous, which in turn encouraged Onondaga County to dump raw sewage into the lake in the belief that Solvay wastes "cleansed" it.

Ernest Solvay's Scientific and Sociological Ambitions

Ernest Solvay gave his name to the large plant in Syracuse and to the town where the plant was located. Solvay himself was never completely satisfied with being known merely as a chemical engineer and business manager. He sought involvement in and recognition from the rarified world of pure scientific investigation. To that end, he became a patron of the sciences and began funding conferences to bring together the greatest scientific minds of his era. He also spent considerable time and effort developing his own scientific theories, believing that it was possible to develop a unified "theory of everything" that included physics, chemistry, and the social sciences. He was convinced that there was a "great universal law" governing all movements and processes that could be applied to the development and organization of human societies, and that it was "possible to declare an energetico-productivist scientific law and to govern the world on the basis of a purely rational order." Unfortunately, while his ambitions involved formulating grand theories, his results were, as Bertraim et al. state, "scattershot," and he often entered into fields of which he had little knowledge. His "gravito-materialitique" theory, which attempted to reconstitute notions of "matter, gravity, and cosmogony," was treated with skepticism by other scientists, who said it lacked rigor. Solvay tended to marshal evidence to support his highly optimistic worldview wherever he could find it, rather than to engage with scientific analysis to see where the process might take him. His attempts to create a kind of perpetual motion machine designed to generate electricity from atmospheric steam ended in a failure that was entirely predictable, given its violation of the first law of thermodynamics.[55]

Still, Solvay's indirect effects as a patron of the sciences were significant. He contributed large sums of money to scientific institutes that he founded. He created the Solvay Institute of Sociology, and hired progressive, liberal Emile Waxweiler to run it. He also hired Waxweiler to organize and run the Solvay Business School, which is still in operation and considered one of Europe's best. Perhaps most importantly, Solvay provided funding for a scientific council, the purpose of which was to explore "current interest in molecular and kinetic theories." The Solvay Council of Physics, which met in 1911, was described by attendee Albert Einstein as a "witches' Sabbath."[56] Here, according to legend, were laid the foundations for quantum theory. The fifth Solvay Council, held in 1927, very famously brought together Albert Einstein and Niels Bohr, and is where Einstein reportedly made his famous quip, "God does not play dice with the universe."[57] Solvay Councils are still held in Brussels at three or five-year intervals. The latest, number twenty-eight, was held in 2022.

Solvay plants were being set up in Europe during a period when it was recognized, by at least some industrialists, that the harsh labor conditions that had gone hand-in-hand with the industrial revolution were likely unsustainable. The rise of labor movements and social democratic political parties, not to mention communists and socialists, was met in some cases with attempts from both political and business leaders to improve working conditions. Otto von Bismarck helped to organize the first system of social security in Germany, not because he was a radical, but because he was politically astute enough to recognize that such a system would help to fend off support for a more revolutionary restructuring of the capitalist system. This kind of reform was not designed to empower workers with control over industrial production, but to treat them better within the system that existed. It was essentially paternalist or "neo-feudal."[58] Solvay plants in Europe were often organized as company towns, designed to foster social cohesion and religious commitments, prioritize family life, and shelter workers from the potentially destabilizing effects of capitalism, while simultaneously channeling their energies and values toward useful, continuous, and sometimes dangerous labor.

While Solvay expressed interest in the ideas of Auguste Comte and Henri de Saint-Simon, social thinkers who supported rationalism and centralized rule by a scientific elite, his company's practices were more in line with the conservative philosophy of Frédéric Le Play, who placed family and community at the center of a pessimistic social science. Le Play proposed that history wobbled between periods of prosperity and economic decline. The former created the conditions that softened a nation's character and undermined moral values, resulting in weaknesses in the social body, which in turn, led to collapse. Collapse fostered discipline, resulting in a hardening of resolve that spurred the entire cycle again, to be repeated *ad infinitum*. Whatever social theories may have animated Solvay, his biographer, Maxime Rapaille, praised the industrialist, contending that he "worked constantly for better conditions for all those who worked in his factories, including the humblest."[59] But since Solvay did not directly manage his European facilities, his chief contribution may have been to establish a humane atmosphere, the specifics of which were implemented by company managers, each according to the conditions at their local facilities.[60] Rapaille also lauded Solvay, claiming that he "loved Nature," and that "only the beauty of his garden touched him."[61] Whether Rapaille appreciated the grim irony of this remark, given all that would follow, is hard to determine, but probably unlikely.

Solvay President Rowland Hazard came from a family that had established a benevolent paternalism in its textile operations in Rhode Island, and he brought that same worldview to the Solvay plant on Onondaga Lake.[62] The Hazards were one of Rhode Island's oldest European families, having settled in the state in 1635. They made a fortune in textiles and were the first company in the U.S. to engage in profit sharing with employees. Rowland Hazard's marriage to Dora Sedgwick, the daughter of Charles Sedgwick, who had been a prominent abolitionist before the Civil War, further reinforced a commitment to social awareness. Dora was an advocate of family planning, women's suffrage, and civil rights for African Americans. She established the Solvay Guild in Solvay, to provide for the basic education, public health, and cooking classes for employees.[63] Nursery services were provided for the wives of the mostly male Solvay employees, where they could leave their children to be looked after at a cost of five cents a day per child.[64] The company held social events, such as balls and picnics, to which production workers were invited. Guild bazaars were organized around themes, such as a Belgian Market Party or a Pioneer Party, and attendees dressed accordingly. During the annual Christmas event, special trains ran from the city of Syracuse to accommodate the large numbers of people who wanted to attend.[65]

Solvay, New York, developed as a company town where small businesses flourished across the street from the plant, where single family homes were constructed on the hill overlooking the production facility, and where workers did physically demanding and sometimes dangerous labor under extremely difficult conditions.

Life in a Company Town

Onondaga County was one of central New York's great recreational getaways beginning in the late nineteenth century. Numerous resorts were built along its shores, which helped to cement the area's reputation as the "Catskills of Central New York," though it became slowly eroded by pollution from the Solvay process and other manufacturing facilities, such as Crucible Steel. There was no official policy decision to privilege industry over recreation and tourism, but the two were ultimately incompatible in such an enclosed geographical space. While industrial concerns were not hostile in principle to recreational activities, and managers and workers in Solvay's industries may even have taken advantage of them, they were, in the end, fundamentally irreconcilable, so the industry won out. The Onondaga resorts did not have the economic clout to challenge the area's industries. Moreover, no meaningful authority existed for them to appeal to.

At the southern end of Onondaga Lake stood the Iron Pier. Completed in 1890, it stretched out along the shoreline, providing a birds-eye view of the entire lake to the north. The pier featured musical performances, dances, circus acts, and even games played by a semi-professional baseball team. An 1899 lithograph depicts an elegant structure hugging the shoreline, with American flags fluttering from the three domed spires on its roof. The waters in front of it are populated with boats of all varieties and sizes, from small sailing skiffs to large steamboats. In fact, steamboats left the pier on a regular basis to visit the resorts to the north and west. The pier was the jumping-off point to experience other sites along the shoreline, while also providing opportunities for fishing and swimming from its various docks. In the image, romanticized naturalism mixes with human activity and energy. Farther to the south, up in the hills surrounding the lake, factories spew out smoke from their stacks. They are not on the lakeshore itself, but their presence in the image indicates how the industrial and the natural were seen as compatible, if not entirely integrated with one another. The factories prefigure the intrusions that would eventually drive the Iron Pier and other major resort facilities from the shores of Onondaga Lake. Not shown in the picture were the hundreds of solar salt vats that abutted next to them, on land that was eventually bought by the Syracuse Street Railway Company. Solvay process waste was even used as fill to build up the area between the pier and the shoreline. Eventually, the pier went out of business when a trolley service eliminated the need for catching a steamer farther up the lake with other resorts as a destination.[66]

The gem of Onondaga Lake resorts was undoubtedly White City. Opened in 1906, relatively late in comparison to some others, it has been described by resort historian Donald H. Thompson as a "spectacular amusement park." White City was a testament to early twentieth-century consumerist excess and a nod to the expanding middle classes that had money, time, and a need for community and connection. The dance floor had space for 1,000 people, surrounded by a 5000-seat open air venue for bands and variety acts. The space also featured a Japanese tea garden, a miniature railway, and even its own small lake, carved into the larger shoreline of Onondaga. Opening day drew more than 41,000 people for a fireworks display, musical performances, and a chance to ride on one of the more than thirty amusement park rides. The most popular ongoing attraction at White City was undoubtedly the Shoot-the-Chutes ride, which featured a long ramp down which boats slid into a lagoon created primarily for this purpose. White City, in spite of its immediate popularity, was shuttered by 1909. The short period between when it opened and closed were the three years when Onondaga Lake began the change from being a relatively pristine body of water to its future status as the most polluted lake in America. White City was on the western shore, very close to the Solvay Process, so it was among the first of the resorts to be severely affected by Solvay's dumping practices. Moreover, during this period of time, automobile transportation was becoming increasingly accessible to ordinary people, so the citizens of Syracuse were able to drive to cleaner lakes such as Oneida or Skaneateles.[67]

Numerous other resorts dotted the landscape on the western side of the lake. The Syracuse Yacht Club, located right across from Crucible Steel, from which steam-powered yachts carried passengers out onto the lake and to various other sites along its shores, and whose regattas became an important social event. While the main building rested on the shore, an elegant clubhouse hovered over the water itself, resting on stilts. More than two thousand people from Syracuse's most elite circles were members of the club. They lunched and dined while watching one of the many regattas that originated from the club's docks, where 150 various sailing and motorboats were moored. The Yacht Club opened in 1898 but closed in 1917 after its main building burned down, and given the increasing invasion of Solvay wastes into its surrounding waters, rebuilding was deemed impractical.[68]

Other resorts included Lake View Point Resort, which rested on a small peninsula that jutted out from the western shore of the lake. It was the first resort to open in 1872 but closed in 1915. Lake View's reputation was somewhat troubled, because the point itself was known as a gathering place for local drunks, who would often engage in fistfights. Lake View was also an early casualty of Solvay waste, given its proximity to the plant.[69] Another resort, Cowan's Grove, which opened in 1874, eventually renamed Pleasant Beach Resort, lasted much longer, closing in 1954. When it opened, it became known for serving a local delicacy, whitefish, which was caught in Onondaga Lake, but by the time it closed, including locally caught fish on the menu was unthinkable, and the whitefish population had been significantly diminished if not eliminated entirely.[70] Smaller resorts included Manhattan Beach, which had a brief run, but was closed at the turn of the century, and Rockaway Beach, which early on was associated with iceboating but then later became known for gambling and serving illegal alcohol during prohibition. Rockaway Beach operated through the 1930s, at which point it was converted into apartments, and was finally bulldozed in 1954 to make way for Interstate 690.[71]

The amusement park, Long Branch Park, opened in 1882 and closed in 1938. It featured a roller coaster, a Ferris wheel, dodgems, a Russian toboggan, and an aerial swing. Across the top of the entryway was the park's motto, "Let All Who Enter Here Leave Care Behind," possibly an ironic take on Dante's injunction at the entrance to Hell, "Abandon hope all ye who enter here."[72]

An exception to the resorts that did not survive the contamination of Onondaga Lake was the New York State Fairgrounds, which were also located near its shores of Onondaga Lake. What saved the site was its distance several hundred yards from the lake itself. It was therefore not subjected to contamination problems in quite the same way as the other establishments. The Fairgrounds opened in 1890 and remains open today. Ironically, not only were the Fairgrounds' continued operations unaffected by the toxic contamination resulting from the Solvay process, one of its parking lots is constructed on one of the Solvay wastebeds.[73]

The pollution of Onondaga Lake would certainly not have happened without the Solvay Process in operation. And the process would not have been possible without the workers who ran the operation. Those workers, along with their families, the small businesses they supported, and the taxes that they paid, created networks of social relationships and economic interchange. They created and maintained the Village of Solvay. It is important that we not consider the lake's pollution in isolation but situate it in relation to the social and economic benefits that arose from it. This is not to justify the actions of Allied Chemical, to minimize the damage to a sacred lake, or to argue for a simplistic cost-benefit calculation. But examining the social and economic forces at play is essential for a full understanding the trajectory of Onondaga Lake's despoliation.

Social Life and Labor Relations in the Village of Solvay

Available historical analyses all suggest that the Solvay Process Company treated its workers humanely, and they responded with hard work and devotion to the company. In her social history of the Village of Solvay, *Smokestacks Allegro*, Rita Cominolli charts a process of immigration and assimilation, as an influx of workers from the Tyrolean region of northern Italy, on the Austrian border, came to the U.S. to work at the Solvay Plant. As with many immigrant groups, these workers established chains of connection that resulted in the establishment of coherent ethnic groups that maintained a strong sense of identity with their European roots, but inevitably were also Americanized in various ways.[74]

The Village was divided along class and ethnic lines. Away from the impacts of localized noise and pollution were the company's executives, engineers, and chemists. They lived in "an enclave of substantial homes" in an area known as the "Uplands." They tended to be WASPs and graduates of top-notch universities, often Ivy League schools. They were innovative in their fields, part of a rising scientific-technical elite, and well-compensated by the company, not only financially, but with recognition and appreciation of their status. The Solvay Process was, in the words of one chemist who left Eastman Kodak in Rochester for Solvay, "very generous."[75] For a member of the elite echelon, it was as good a place to work as anywhere else in the U.S.

If Ernest Solvay supported principles of paternalistic support for workers, partly out of a belief in social obligation, but also as a means to stave off worker rebellion, those principles were adopted and intensified in the American context, as part of the managerial philosophy of Frederick Roland Hazard. Solvay employees, like their managerial counterparts, were well-compensated financially. But beyond that, Hazard took a paternalistic interest in many aspects of village life. The Hazards were, according to Cominolli, "making a sincere attempt to treat workers fairly,"[76] and the Solvay Company was an early adopter of the eight-hour workday. The company established a food service, charging employees seventeen cents for meat, potatoes, and soup, sourcing the food from land the company had bought in Tully from which salt was mined. It established a "penny savings" bank for employees (a bank which encouraged employees to make small savings contributions every month), as well as health insurance and a pension plan. Perhaps most significantly, it created a generous profit-sharing program that, during flush years, provided some employees financial benefits that matched their salaries. The Hazards were also prolific supporters of various community charities, such as the YMCA and the Community Chest (the precursor to the United Way).[77]

Due to this benevolence, Solvay Process and Allied Chemical, in the early years at least, experienced an absence of labor unrest at a time when labor conflict, at such firms as Ford Motor and the Pullman Companies, was becoming increasingly violent. Things were so quiet at Solvay Process that the Labor Department approached company executives for advice on how its principles might be applied more widely.[78] Labor conditions at the plants were difficult and dangerous, but, because managers were believed to be fair in terms of compensation and benefits, no serious attempts to unionize were made through the early decades of the company's operations.[79]

Perhaps the best source we have of what life was like in Solvay in the early to mid-twentieth century is collected in two slim but remarkable volumes of testimonials: *Solvay Stories*[80] and *Solvay Stories II*.[81] The project was undertaken by Judith LaManna Rivette, a Solvay native, who spent her life in the area and became a successful attorney. The two books provide a glimpse into the culture of a small factory town during a period that ran parallel to Onondaga Lake's contamination.

Rivette depicts an attractive world of hard-working immigrants who created strong bonds of community. Life revolved around work, family, food, education, church, and civic participation. Along with the Italian immigrants, there were also Germans, Swiss, and a smattering of other ethnic groups. The Italians were devout Catholics, and St. Cecilia's, the village's Catholic Church, was a hub of the community activity. The town also had Protestant churches, including the Asbury Methodist Church, which was the largest.

The village had many small shops and stores that provided the necessities for members of the community. Rivette depicts shop owners as often colorful, generous figures, who gave candy to children and extended credit to families in need when times were rough. The town also had many taverns, which, besides the churches, were central to social interaction. The taverns, in turn, supported softball teams and other amateur sports clubs. Opening a small business in Solvay provided an opportunity to move out of one of the local factories to an economic activity less dangerous and dirty, and with more autonomy. But it didn't necessarily prove to be an improvement in economic circumstances, given the precarious nature of small business.

As someone who grew up in a factory town in Western, New York, I saw parallels to my own childhood as I was reading the accounts of life in Solvay. Like Solvay, Jamestown was in the snow belt (the result of lake-effect storms blown in from Lake Erie in Jamestown's case, Ontario in Solvay's case). Kids sledded down Solvay's snowy hills, with lookouts at the bottom to warn of oncoming traffic, given that the end of the run was one of the village's main streets. As in Jamestown, Solvay's fire department flooded tennis courts in the winter to create makeshift skating rinks, providing another winter activity for the community's young people. Snowball fights were, of course ubiquitous, as was the almost daily chore of shoveling sidewalks and driveways.

The Solvay Process was the engine that provided the economic support for the community. Working in it was difficult. Workers were surrounded by toxic substances and other dangers associated with any large industrial enterprise, which no doubt shortened the lives of many, but workers took pride in their work and were relatively well-treated by the company. Solvay Village was not a classic company town where the housing, most of the stores, the banks, and nearly all economic sectors are owned by the company. As noted earlier, in Solvay, independent businesses were woven into the community fabric.

The workers collectively bargained under the umbrella of a local branch of the Chemical Workers Union, but workers also recognized that the town's existence depended on the continued operation and profitability of the Process. In exchange for labor and community loyalty, the company yielded numerous benefits to its workers, not just decent wages, but vacations, health insurance, and the promise of a pension after twenty-five or thirty years on the job. Since Allied had its own electrical facility, it provided electricity to the entire community at exceptionally low rates, a fact that further cemented bonds of loyalty to the company.

The Process was not a distant entity to which workers commuted and then left for distant suburbs once they clocked off the job. Allied was, in physical terms, in the very heart of the village. On one side of the main thoroughfare, Milton Avenue, were the various production facilities. On the other side were houses that extended southward up along a steep hill. Solvay's famous Stone Pile, a mound of limestone about sixty feet high, which served as a reserve stockpile for the plant, sat within a residential neighborhood. It was a prominent geological feature and was viewed favorably as a local landmark, if it was even noticed at all. The Solvay works themselves were an ever-present aesthetic backdrop to the community. One of the more noteworthy aspects of Rivette's collection is the lack of attention that's given to living within close proximity to an unattractive, smelly, and noisy industrial district. Local residents seemed focused on the social life of the community and their commitments to one another, rather than to what would likely be considered a nuisance to strangers or newcomers to the area. For Solvay residents, the Process and its ancillary works were simply a fact of life, and a source of community pride.

Solvay was not the only industrial concern in the area. Crucible Steel, which also generated a fair amount of air and water pollution, was the second largest employer. Along with it, there was Iroquois China, which made durable, well-designed porcelain affordable to middle-class consumers. Ironically, the natural world was integrated into the community's fabric to a greater extent than is currently the case in many suburban enclaves. Even though industrial operations were in close proximity, residents carried practices from their European heritage. Many people had gardens, especially during World War II, but before and after as well. Residents grew grapes and made homemade wine that they provided at community events. Some grew corn, which they made into homemade popcorn. People would pick raspberries in a patch that was nearly continuous with the Stone Pile, and sheep grazed on the Village hillsides through the 1940s.

Fast food was virtually unknown. Culinary traditions were also brought from Italy and other European countries, which included healthy servings of vegetables, such as eggplants and tomatoes, and peasant foods such as polenta. Diets were thus varied, and not focused entirely on what became the typical American diet of meat and potatoes. And people walked. Especially early in the twentieth century, automobiles were a scarce commodity, so walking was the most common form of transportation. Given Solvay's hills, walking was sometimes strenuous, so, while exercise may not have been a part of any conscious plan, it was a necessity, another simple fact of life.

The various refuse piles in the community became a source for recycling. Kids would go to industrial landfills to find usable products. Iroquois China threw its broken dishes out in the back, which would be carefully sorted through by the area's children, who would sometimes find still intact and usable items, which they could sell or bring home to their families. The local paper company threw out large amounts of waste that became paper that students could use for their needs at school. While much waste was generated by the local factories, what was usable was often used. However, much of the waste generated, such as the mountains of calcium chloride, had no uses whatsoever. The lines between natural and industrial were sometimes blurred, such as when the Stone Pile became a kind of playground for the area's children, until they were chased away by the Allied security guards. Kids swam in the Erie Canal, which also provided a source of water for some of Solvay's industrial activities.

It should be noted that Solvay's residents were not entirely unaware of the environmental impacts of the Solvay Process. Outsiders, generally residents of the City of Syracuse, considered the Village to be a place that smelled bad, most noticeably from ammonia smell that was a constant presence. Those who lived closest to the plant often had to sweep soot from sidewalks and to shake it out of laundry that had been hung on clotheslines. If Solvay Village residents generally overlooked or had become used to the pollution that marked life there, others did notice it, which made villages residents a bit self-conscious. Perhaps the opprobrium from outsiders, particularly the people in the larger Syracuse Metropolitan area, even strengthened their sense of kinship. Of course, sometimes the contamination problems were so intense that they simply could not be ignored even by residents. In 1943, a dike holding back Solvay waste sludge broke, and 40,000 tons of calcium chloride waste flooded the State Fair Grounds, the Boulevard, and houses northeast of the Village, covering eighty-five acres on the western border.

No doubt the Villagers' healthy lifestyle and the support provided by strong social bonds helped to offset, at least to a small extent, the deleterious effects of living in a polluted environment and made it more tolerable. This was not only the case in Solvay, but also for much of industrial America in the twentieth century. In exchange for economic sustenance and viable communities, incredible amounts of environmental contamination were tolerated, and once industrial activities were started, regardless of their damage to the environment, they were difficult to stop.

As striking as the depictions of the idyllic community life presented in the *Solvay Stories* are, certain absences are noticeable. Diversity in the Village operated within what would now be considered fairly narrow bounds. "Mixed marriages" involved Italians and Poles, or, more daringly, perhaps, between a Catholic and Protestant. No mention is made of African Americans. And religious diversity seems entirely within Christianity itself. Solvay did not have a local synagogue, for example, or a Greek Orthodox Church.

Most striking, perhaps, in the *Solvay Stories*, are the lack of references to Onondaga Lake itself. There are no mentions of the lake's beauty, or even its presence. The Village was on the other side of Solvay Process, not directly on the lakeshore. Still, the lake was only about a mile from the Village center. Despite its proximity, children seemed never to venture over to it. The old Erie Canal and Geddes Creek are mentioned as occasional swimming holes, but never the lake itself. You couldn't swim in the lake, and its nasty smell and lime green color turned it into a nuisance and eyesore. Moreover, it was not among the things that Villagers could point to with pride. More likely it was an object of collective guilt or shame. The community's vitality was entirely dependent upon Solvay Process's ability to dump its wastes into the lake's waters or into the various waste sites that dotted the company's property. Even as the lake's unfortunate reputation as the "most polluted lake in America" grew, the residents of Solvay, at least as represented through the *Solvay Stories*, expressed very little interest in or concern about it. No mention is made of Village residents being involved with initial or later cleanup efforts. For the workers who lived in Solvay, it was as though Onondaga Lake didn't even exist.

Working conditions began to change at Solvay after the 1920 merger. Solvay Process, Semet-Solvay, General Chemical Company (which produced acids), the Barrett Company (which made tar products), and National Aniline and Chemical Company, Inc. (which produced coal tar derivatives and dyes), merged to form the Allied Chemical and Dye Company. At the same time, many of the original founders of the company such as Roland Hazard and other members of the Hazard family died. The new managers believed that the original owners had been too generous with their employees, up to and including the managerial strata. Cominolli states that, "More often than not, the company's new blood wanted to do away with any managerial policies that had been implemented by the Hazard administration." A shift from benevolent paternalism to bureaucratic control and managerial efficiency were introduced into running the company. Allied was the largest chemical company in the U.S., and was highly profitable, even through the

Depression-era 1930s, but management embarked on a policy of stream-lining, what would today be called "downsizing." They "clean[ed] house completely," and, according to one manager "'All they did was fire.'"[82] They put an end to all activities deemed to be "non-productive." They ended the company's profit-sharing program, stopped all welfare work, cut out the gyms, readings, schools, and the day nursery. Many engineers and managers either left or were fired, and they were often able to find jobs at other companies, where they provided skills and knowledge that helped to benefit the company's competitors. Allied's competitors began to establish alkali facilities overseas, in such far-flung places as China and Japan.[83] As a result, companies such as Dupont and Union Carbide challenged Allied's dominance, and by 1930, they had fallen from the largest chemical country in the U.S. to third, in no small part because it became, as one worker put it, a "gloomy place" to work.[84]

PART II

Environmental Movements

3

Pollute the Lake, Save the Lake

The Emergence of Awareness and Activism

Harvey Baldwin, Syracuse's first mayor, had a vision for Onondaga Lake. In 1847, before he was elected mayor, he said of the lake that

> All its beautiful shores will present a view of one continuous villa, or-
> namented with a shady growth of hanging gardens and connected by a
> wide splendid avenue that will encircle the waters and furnish a delightful
> drive to the gay and prosperous citizens of the town who will, toward the
> close of each summer's day, throng to it for pleasure, relaxation, and the
> improvement of health.[1]

But the lake was hardly pristine even then. Bluffs along the eastern side of the lake were occupied by vast salt drying beds, and huge amounts of brine were diverted from them into the southern end of the lake, which had become a "bog and a mosquito infested swamp."[2] In 1893, a roadway was built along the western shore of the lake. The roadbed was made by piping Solvay Process waste and allowing it to harden.

Starting in the late nineteenth century, two broadly competing forces, each with a different agenda, collided over the fate of Onondaga Lake in a competition that continued through most of the twentieth. On one side were people with power and influence, including those who ran Solvay Process, the City of Syracuse, and other industrial concerns. All sought places to deposit their waste products, whether in the lake itself, in the streams that flowed into it—Geddes Creek, Nine Mile Creek, On-ondaga Creek, and Six Mile Creek—or on the surrounding land. On the other side were those forces trying to protect the lake from environmen-tal threats, and to fulfill the vision of Mayor Baldwin and others who saw the lake as being a natural wonder that could provide the area's residents with a place for escape, relaxation, and aesthetic appreciation. The conservationists who wanted to protect the lake had very little clout at first. They were a small minority fighting against the tide of technological and industrial progress. However, over time, their influence grew. The citizens of Onondaga County who sought to prevent further pollution and begin the process of restoring the lake, starting in the 1940s, were on the cutting edge of an environmental movement that would eventually sweep through the U.S. in the 1960s and 70s. While early environmental protectionists were in many ways ahead of their time, their focus tended toward preserving the environment for human use: fishing, boating, swimming, and other recreational activities. Eventually other voices, specifically Native American voices as represented by the Onondaga Nation, contributed to the conversation on the lake's restoration, not just for recreational purposes, or even aesthetic appreciation, but because of its deep cultural significance and sacred character, although their voices did not, in the end, have equal weight.

A Devil's Bargain

In Syracuse, as in all American cities in the nineteenth century, sewage disposal was a haphazard affair. Human waste from the expanding city was flushed into open gutters, thrown into the streets, or, at best, piped into the nearest body of moving water, where it was hoped that the sewage would eventually be washed away. Onondaga Creek, which runs through the center of Syracuse into Onondaga Lake served as a convenient site for this purpose. The creek became increasingly fouled as the city's population grew, with a stench that intensified during the warmer summer months, when the water would slow to a trickle and live up to its unfortunate moniker, "Shit Creek." Starting in the late nineteenth century, in response to citizen complaints and public health concerns, Syracuse, like other cities, created a vast underground sewer system which transported wastes away from urban neighborhoods, partially bypassing its creeks, piping it directly into Onondaga Lake. As with other cities, the Syracuse system combined household sewage with storm runoff. While such a combined system would eventually create significant problems for waste management, at the time it made sense, given the dependence of sewage disposal on dilution. More water during rainy periods would help to wash away the smelly, nasty, disease bearing, human wastes that accumulated in the drainage systems.

Many large cities in the northeast and Midwest had broad, fast-moving rivers or large deep lakes nearby, into which sewage and other wastes could be deposited. Buffalo, Cleveland, and Chicago were located on Lakes Ontario, Erie, and Michigan respectively. Pittsburgh was at the confluence of the Alleghany and Monongahela Rivers. Eastern coastal cities had the Atlantic Ocean. The negative impacts of continuing waste disposal on these large bodies of water could be denied or ignored, at least for a while, because of their capacity for absorption and dilution. Onondaga Lake was small by comparison, with a relatively slow flush rate. Thus, the dumping of sewage from Syracuse and nearby towns into Onondaga Lake had a noticeable impact almost immediately.

In 1896, the City of Syracuse lodged a complaint against Solvay Process for dumping its waste into what had until that time been a relatively clear, unpolluted glacial lake. The process created large amounts of calcium chloride left over from the spent limestone from which the carbon, necessary for soda ash production, had been removed. While the substance was inert, and not in itself toxic, it was not entirely benign, given its contamination with other chemical substances on its way to disposal. And there was a lot of it, tons and tons of it. At first, it was dumped onto land owned by the company, but the quantities were so vast this soon proved to be insufficient.

The city had its own problems with sewage waste disposal. Dumping raw sewage into Onondaga Lake was obviously fouling it and causing public concern during a period of time when it was still an active site for recreational purposes. In the late 1890s, major fish die-offs were becoming common, likely caused by algae blooms and bacterial infestation. The lake's decline occurred during a period when resorts were springing up, primarily on its northern and eastern shorelines. In response to their mutual interest in finding solutions to their waste disposal problems, the city and the company reached an agreement. Solvay Process could dump its solid waste near city owned parts of the lakeshore, and, in return, the city would be able to deposit its sewage into wastebeds, sequestered with dikes that had been constructed by the company to the west of the lake near Nine Mile Creek.[3] The Solvay wastebeds seemed like an ideal solution, because the wastebeds had no other useful purpose, and the calcium chloride reacted with the sewage to mitigate its smell.

In November 1911, Eugene H. Porter, the State Commissioner of Health, ordered that the Village of Solvay and the Solvay Process company construct a plant to stop the pollution of Onondaga Lake. The Commissioner's order covered both sewage and chemical contamination, saying "The waters of Onondaga Lake are polluted by the sewage of the city of Syracuse, by the sewage of the village of Solvay, and by the chemical waste of the Solvay Process company. This pollution, especially near the mouths of Onondaga Creek and of Harbor Creek is very marked and constitutes a public nuisance."[4] Through a spokesperson, the company indicated that it would comply with any state orders, but there is no indication that it did, or even attempted to.

The city and Solvay Process continued polluting the lake and its environs for years, until in 1921 the city formalized its agreement with the Solvay Company and sold it ninety-one acres of land within the lake bed itself, on which the company agreed to construct a set of docks and other "structures," to facilitate use of the lake by area residents. As Solvay waste materials continued to pour into the lake, city leaders seemed to believe that the ledge that was being created could be useful to facilitate recreational use of the lake. While this may seem questionable in retrospect, the company failed to follow through, and its inaction would eventually become a source of friction between the city and the company.[5]

The city's agreement with Solvay Process coincided with its movement toward the construction of its first sewage disposal plant. An early proposal for installation of an Imhoff tank was forestalled when the U.S. Congress refused issuance of bonds to the city. The Imhoff system, named for the German engineer Karl Imhoff, consists of two tanks. As sewage flows through the top tank, it is aerated, with heavier elements falling into a second tank, where they are subjected to bacterial digestion in a process that can take months to complete. Versions of Imhoff tanks are still in use today, and later editions of his classic work, *Handbook of Urban Drainage*, are still in print.[6] The tank was never constructed, but the land designated for it was purchased by the city on Hiawatha Boulevard, at the southern end of the lake, near the mouth of Onondaga Creek.

In 1924 the city completed construction of a Dorr clarifying system, consisting of four large cement settling tanks, with skimmers. As with the Imhoff tank, sludge collects at the bottom, but it is quickly removed, leaving little time for anaerobic digestion. The cost of this system was considerably less than the earlier proposal, saving the city $1.5 million (equivalent to $29 million in 2021 dollars). At first, New York State was reluctant to sign off, but a consulting firm hired by the city to review the process gave its approval, ultimately convincing state authorities. The consultants, Metcalf and Eddy, deemed the system "practical," contending that it would "prove adequate for twenty years" and perhaps "much longer." The "Imhoff tanks" and "trickling filters" were unnecessary, they argued, because, "Disposing of the sludge by mixing it with industrial wastes . . . will prove highly effective."[7] In other words, the engineering consultants approved a system that would send bacterially active sewage sludge into the Solvay wastebeds for "sterilization" as an alternative to using proven methods of primary sewage treatment. The sludge would be sent through a small pipe, two-and-a-half miles long. Since such a process had not been previously attempted, operators started with a highly diluted solution, but over time discovered that "a smaller quantity of more concentrated sludge may be pumped at lower velocities without difficulty."[8]

The system also had a screen and "grit chamber," of particular importance in combined sewage treatment systems, because of the detritus that ends up in storm sewers. The screens collected rags, paper, and "a small amount of fecal matter." The sand and gravel that piled up in the grit chamber was spread over walkways and roads at the facility or buried. A particular nuisance was caused by the scum removed via the skimming process, which consisted of oil and grease that people had thrown down storm sewers, animal carcasses, rotten vegetables, and other "organic matter." Operators learned that once this material was dried out, it could be burned, so that was how they handled it.[9]

A major problem, and one that would continue to haunt Syracuse Metro Area sewage disposal for decades, was caused by overflow. Heavy rain events, or melting spring snows, would flush so much water into the storm drainage system that it became overwhelmed. When this occurred, the raw sewage was allowed to flow directly into the lake.

In May 1928, Walter Ewanski brought the first environmentally-oriented lawsuit against Solvay Process. Ewanski sought $10,000 damages to his business and land over a two-year period at a time when the company began dumping chemicals into Nine Mile Creek near his property. According to Ewanski, no authority had given the company permission to foul the creek, which had been considered public property. The city had deeded the company a strip within the lake itself, on which it had promised to build a dock. But Solvay never completed the structure, leading to a potential legal claim by the city (although the city never followed through). According to the suit, "The mouth of the lake has been clogged by refuse and during the high tide the slippery substance is washed back into the land of Ewanski."[10]

While Ewanski's lawsuit indicates recognition that significant pollutions problems were occurring, the newly established County Commission was moving to create a park on the shores of Onondaga Lake. A visit to the area by members of the Commission and the Civic Development Committee of the Chamber of Commerce resulted in the conclusion that there were "great possibilities both in the beautification of the eastern shore of Onondaga Lake and in the consequent trend of residential building in that direction." The proposed plan included both a parkway and a beach. The "biggest menace" to the plan was of course the pollution problem: "Solvay Process has been using the lake for years under an agreement which protects it, and a considerable part of the old lake has disappeared in the consequent filling process." The other issue was sewage, which the committee agreed could be resolved by the city whenever it decided to do so. But the proponents were optimistic in spite of these obstacles, with one member stating that "All of Syracuse will be glad to have Onondaga Lake restored to the beauty and availability as a pleasure resort that it enjoyed thirty years ago."[11]

The Commission's vision was wildly optimistic and highlighted the contradictions at play in the ecological decline of Onondaga Lake. Community members, even members of the business community, recognized the lake's value in terms of its recreational and potential economic benefits. But the pressures favoring industrial development with profits for owners and jobs for workers were in direct conflict with the perspective of a nascent ecologically sensitive movement. For most of the twentieth century, the former would simply overwhelm the latter. In addition, the

weight of simple inertia cannot be discounted. Once Solvay began operating, without a permit or regulatory oversight, few mechanisms existed
to bring it under control or challenge it in any substantive way.

In the 1930s, partly as a result of Depression-era public works projects, Hiawatha Boulevard was constructed along the eastern shore of the
lake, with an adjoining picnic area and ball field.[12] In retrospect, the city
seemed to be in denial about the lake's deteriorating condition. While
the city was facilitating development projects, it had become increasingly obvious to even the most casual observer, that the lake was unfit
for most recreational uses.

In December 1942, the State Water Pollution Control Board, an agency
set up by the state legislature to classify all of water in the state according
to a standard for use, met to discuss Onondaga Lake. This would be the
first formal evaluation of the lake's ecological condition and would be
followed by dozens of others over the decades that followed. The meeting included a variety of stakeholders, including representatives from
Solvay Process. The board considered options for the lake's use: Was it
fit for industrial purposes, sewage disposal, and/or recreation? Fishing
was of primary concern and was included under the category of recreation. The board determined that conditions were apparently favorable
for "propagation" (given that fish could apparently live in the lake), but
the "fish from the lake are oily to the taste." (Determining whether the
fishing standard had been met apparently involved some unfortunate
soul having to actually eat, or at least taste, the fish, in an era before testing protocols were well-established.)

An engineer from the state health department wisely nixed the idea
of using the lake for drinking water, largely due to the raw sewage that
was being dumped into it. Indeed, at this point the lake had become
so polluted that it was unfit even for industrial uses. Williams B. Kin-
slow, a representative of the Crucible Steel Company, noted that their
factory was using "considerable quantities" of water from the lake, but
that pollution levels were causing problems for them. In fact, they were
considering moving their inflow pipe from one part of the lake to another to avoid the intake of sewage. When remediation was requested,
representatives from the city were noncommittal, suggesting that before acting, they would need to know the costs of reducing sewage outflows. The city's sewage engineer, Joseph Keiffer, perhaps in an attempt

to deflect criticism from sewage treatment, focused his attention on oil discharge from barges in the channel and the emission of fly ash from trains that was landing on the surface of the lake. What was abundantly clear from the meeting was that Onondaga Lake faced multiple, complex, and intertwined contamination threats, so many, in fact, that it was possible for one party to try to avoid blame by pointing a finger at another, who it could reasonably be argued, was doing something even more egregious.[13]

Attempts were made in the 1940s to contain both the chemical contamination and the sewage by using dikes, but this was ultimately unsuccessful. In 1944, a "goo flood" made the extent of contamination evident, as untreated waste spilled out of the dikes and into the lake in a very obvious and horrifying way. This seemed to represent a kind of turning point, as civic leaders began to challenge Solvay itself for despoilment of the lake. William Maloney, an early and consistent advocate for the lake, took the company to task at a Chamber of Commerce meeting, at which he accused Solvay of committing the "crime of the century" against the lake. Maloney said he had contacted the State Attorney General's office about the problem and had received word that they would look into it. As Maloney knew and stated, however, the city was itself complicit, given the amount of sewage that it was dumping into the lake. The synergistic relationship between the city and Solvay, between public and private, would be an ongoing problem in any attempts at a cleanup.[14]

In spite of the presence of the Dorr clarifier, 60% of the sewage going into the lake was completely untreated, with the rest only partially treated. The system was simply inadequate to meet the needs of a growing metro area, and the combination of sewer and storm drains continued to generate overflows during periods of snow melt or heavy rain. Nine Mile Creek became the conduit by which Solvay's contribution was finding its way into the lake where "at times," according to the *Herald American*, "the stream runs almost milk white far unto the lake with Solvay waste[.] For several hundred yards out into the lake at the mouth of the creek, bubbles arise constantly as in a stagnant pool, indicating waste is decomposing without sufficient oxygen." But the Solvay wastes were not decomposing, because the calcium chloride that was being dumped into the lake was a highly stable chemical compound. It would pile up at the bottom of the lake in huge mounds, which, unless eventually removed, would remain through a period of time measured in geological epochs. Dikes in front of the Halstead Steel plant contributed to the Six Mile Creek outflows, "feeding a gooey mess" that moves under State Fair Boulevard at a rate of around four miles per hour.[15]

Coalescing Forces Press for a Cleaner Lake

In 1946 cleanup efforts began "gathering momentum" as "citizen-inspired movements to return Onondaga Lake to its natural state" began to coalesce, bringing the involvement of New York State officials along the way.[16] It should be noted that national environmental movements at this point were primarily focused on wilderness preservation, so for a community to address and challenge industrial pollution in a direct way, was relatively rare. The New York State Legislature even became involved. The State Conservation Department (before it was renamed the Department of Environmental Conservation) agreed, based upon prodding from the Liverpool Chamber of Commerce, to send experts to do a survey of the lake pollution problem.[17] Liverpool is on the opposite side of the lake to Solvay, so residents there were receiving few of the economic benefits associated with the lake's contamination, while experiencing significant negative impacts.

The state also took action to recover the ninety acres of land that had been deeded to the company at the turn of the century.[18] Allied had never built the structures as promised, and the deeding of the land apparently became the rationale for sending effluent into the lake. Moreover, there was significant encroachment of effluents beyond the ninety acres, so the Solvay Process Company, had, in effect, violated the county's property rights by extending its contamination out into larger and larger parts of the lake.[19]

The Conservation Department began its survey in late June of 1946, by analyzing the lake's fish population. After pulling several dozen carp from the lake, the investigators concluded that "pollution in Onondaga Lake has affected only the quality—not the quantity—of fish life." While carp admittedly might not be the best indicator of the lake's health given their ability to withstand contamination, a number of yellow pike were also caught. The survey's director, U. B. Stone, said that they were probably okay to eat "if they were well-cooked to kill the bacteria." He went so far as to suggest that he'd probably eat one himself but might be disappointed because of the "taint." The disappearance of whitefish, he noted, could be due to a number of facts, but he didn't specify them. All fish caught were to be analyzed further in laboratory settings. Tests were also conducted of chemical contamination in the lake which pointed to the presence of "phenols, ammonia compounds, and sulfides." Oxygen levels were also tested.[20]

State Senator Richard P. Byrnes and Assembly member Lawrence M. Rulison proposed a resolution to set up a committee to investigate the pollution of Onondaga Lake and to propose appropriate legislation to address it. Since the resolution was proposed too late in the session, however, no vote was taken.[21]

Maloney had become the leader most associated with challenging Solvay's activities, and the Liverpool Chamber of Commerce became the organization most associated with pushing the state to address the pollution problem. While it might seem odd that a business-friendly group like the Chamber of Commerce was taking the lead in an environmental fight, Liverpool businesses were being threatened, while gaining virtually nothing in return.

The construction of the Onondaga Lake Parkway, a roadway facilitating access to the eastern shore of the lake, helped to spur the cleanup movement. The parkway not only brought people along the lakeshore on a routine basis, but its construction was completed in tandem with a new lakefront park. Easier access to the lake made the extent of its pollution and the smell associated with it obvious to a larger segment of the area's residents. The *Syracuse Herald-Journal* now began to editorialize in favor of the lake's protection, stating categorically that "every practice that despoils the lake should be stopped." The same editorial stated that the pollution should never have been allowed to begin with and that "Any city with a natural asset like Onondaga Lake should make the most of it and not abuse it."[22]

The number of interested community members kept expanding. A group of local emissaries took an informal boat tour of what they referred to as the "alleged" pollution problem in the lake that summer. The tour was arranged by the president of a local bank and included both Republican and Democratic politicians, the State Fair Commissioner, an apple grower, a parks board member, and a member of the local anglers' association.[23] The party's boat apparently became stranded in the polluted mess on the Solvay side of the lake, and someone on board joked about it: "We'd better get out of here or Solvay Process will start taking pot shots at us." One businessman, Frank Sawmiller, noted the difference in the lake from when he was growing up, when people used to get drinking water from it. Sawmiller's family had a business cutting ice that they sold for packing fish, but they had to move it to Lake Oneida as the pollution intensified. One member of the party described the pollution as "a rank imposition on the public."[24]

Allied Chemical began to fight back. It claimed, for example, that a statute of limitations applied to the state's claim that it controlled the ninety acres that had been deeded to Solvay Process for the building of docks or other various improvements to the lakefront. Maloney met with Assistant Attorney General Arthur Mattson to argue that it did not. In fact, Maloney noted, the deed stipulated that the state had the option to buy the land back at the same price they sold it. The Attorney General was noncommittal, merely stating that his office had the public interest at heart, and that the case had become complicated by the number of competing investigations and charges, which at this point included a State Health Department investigation, state legislative investigations, and the Governor's Committee on the State Fairgrounds.[25]

G. A. Holmquest and Earl Devendorf of the State Health Department began another assessment of the lake, a prelude to a more complete analysis that would be forthcoming.[26] The *Herald Journal* again demanded cleanup of the lake, while simultaneously recognizing the importance of the Solvay Company as an economic asset to the community. While expressing support for Solvay in general terms, it also stated that "further lake pollution *must be prevented*."[27] (Emphasis in original.)

Shortly after the Health Department assessment, a hearing was held in Syracuse by the Joint Legislative Committee on Interstate Cooperation. Approximately forty residents turned up for a spirited session in which the chair asked that there be no name calling but "few speakers failed to heed the stipulation." Maloney, in what would turn out to be a prescient comment, said, "It has taken 50 years to get the lake in its present polluted condition: we sincerely hope it won't take another 50 years to clean it up." The company was represented by John P. Burns, who made no apologies or commitments. He contended that "the people who want to clean up Onondaga Lake in a hurry are too ascetic-minded. They forget there has to be industry so people can eat."[28] (He probably meant aes*th*etic-minded.)

Due to the pollution, particularly the offensive sludge piles, a move was now afoot to move the New York State Fairgrounds, located a few hundred yards from the western shore of the lake, to another location.[29] The proposal to move the fair to what was at the time known as the Mattydale Army Air base, was met with voluble opposition. Opponents feared that moving the fairgrounds would allow the Solvay Company to expand the area where it was dumping its waste and make the problem even worse. Assemblyman Cleland Forsythe noted, "There has been considerable fear expressed by some people . . . that should the fair be moved from its present location, the grounds might be put to use in handling still more of the Solvay Process Company's waste." Governor Thomas Dewey was drawn into the matter, and he ordered the state Conservation and Health Departments to complete surveys of the lake.[30]

Things seemed to quiet down a bit in the first part of 1947, but in the fall a worker accidentally released a large quantity of waste into the lake from the dikes that were holding it, bringing renewed attention to the issue.[31] At the same time, the City of Syracuse, the Village of Solvay, and Solvay Process all declined to participate in hearings held by the County Pollution Commission, much to the chagrin of chair Don H. Brown.[32] The Commission's hearings were otherwise well attended, and included William A. Maloney, leader of what was now termed the Onondaga Reclamation Association, Inc., and Dr. Charles Alvord, who represented the Onondaga County Federation of Sportsman's Club. The Mayor of Liverpool was also there, and stated, "It's a crime when you see what [Onondaga Lake] is like today."[33] Maloney was becoming increasingly frustrated with New York State and the City of Syracuse's unwillingness to address the lake's water quality issues. He even accused the State Commissioner of Health, Dr. Herman E. Hilleboe, of being lax in his duties to protect the public's health. He asserted that the company was in violation of its permit, and that the state had to power to curtail discharges under its Public Health Law.[34]

Perhaps due to mounting public pressure, Hilleboe soon announced that the state had initiated negotiations with the company, and that if these failed to produce results, he was open to pursuing legal action against it. Hilleboe said that he had discussed the matter with Deputy Attorney General Mattson, who had indicated current New York State statutory authority provided the means to bring both the company and the city into court. An air of confidence seemed to pervade the statements of state officials, as though they believed that it was only a matter of time, and not that long a time, before sewage from the city and chemicals from Solvay were no longer flowing unimpeded into Onondaga Lake.[35]

Businesses, at least some of them, continued to coalesce behind the push for lake cleanup. Crandall Melvin, President of the Merchants National Bank and Trust Company, wrote an editorial in support of water pollution control legislation. He suggested that there was a $100 million annual loss to the local area from water pollution, and made a pitch especially for the building of wastewater treatment plants, not only in Onondaga County, but across the nation as a whole. In this way, he was expressing support for the Federal Water Pollution Control Act of 1948, which was the first piece of national legislation to provide some federal support for the construction or improvement of municipal sewage systems. He noted that the federal government had ordered Solvay Process to stop polluting Onondaga Lake as early as 1907, but with no obvious effect.[36] Melvin had been involved with the lake cleanup issue from the start of the recent campaign, and the voice of an important business leader added credibility to calls to move it forward.

An op-ed entitled "Authors of Pollution Shirk Guilt" seemed to capture people's mood of frustration. Author Grace Lewis opened her piece with the statement, "Despite official pronouncements that Onondaga Lake may become a serious health menace at any time and obvious evidence that the small body of water is a virtual cesspool of infection, some direct contributors to the abominable condition are still 'playing ostrich' and refusing to accept responsibility." Lewis contended that the city of Syracuse and Onondaga County were taking steps to address their sewage discharges into the lake, but that the Village of Solvay and Solvay Process were dragging their feet. When asked about the Village's responses to what was increasingly being cast as a disease-infested cesspool, the village attorney refused to comment, suggesting that he was "only the village attorney and not familiar with the subject matter." Company spokesman J. H. Hahn also gave no comment on the specifics but stated that "Solvay Process has always been very civic minded . . . I'm sure at the appropriate time Solvay will play along with the public and with the community." At the same time, surveys of the lake continued to multiply, and a group of Syracuse University professors had been granted $46,000 to complete an analysis of the entire lake, including all of its tributaries.[37] Some members of the community were suggesting that Governor Dewey's stance on the lake cleanup should be considered when casting a vote in the upcoming presidential election.[38]

Water pollution was not the only problem that community members were focusing on, but a different constituency, mostly residents from the Village of Solvay itself, were taking the lead on air pollution issues. This made sense, given that they were the ones experiencing its most direct impacts. The Solvay Air Pollution Committee held its first meeting at the Solvay Village Hall in September 1948. Residents of the area reported recent increases in "smoke, lime dust, fly ash, and fumes." The discharges were so bad that children were kept inside, windows had to be kept closed, a white cake was said to have turned completely black after being put out on a table for five minutes, and neighbors could not see one another's nearby houses, because of the dust.[39] One Army vet who had been a "gas NCO" during World War I said that he had detected the smell of chlorine gas in the air.[40]

Syracuse had a smoke abatement ordinance on the books, and some modest moves were made to enforce it, or at least to suggest that it should be enforced. But as Mayor Frank J. Costello noted, the quality of the city's air was in large measure dependent upon factories in the area, such as the Solvay plant, that were outside of the city's jurisdiction.[41] The *Syracuse Herald-Journal* editorialized in support of the residents of Solvay in their demands for cleaner air, going so far as to support a county-wide smoke abatement ordinance, which would give local authorities legal leverage against the Solvay plant. But it also commended the company for its proposal to take steps (unspecified) to resolve the air pollution issue.[42] A week later, another hearing was held, this time seeking testimony from residents who lived somewhat farther from the Solvay facility. They also complained of significant air pollution, including health effects and property damage (such as the ruining of the finishes of their automobiles). Residents suggested that their property values were being negatively impacted and spoke in support of the smoke abatement proposal.[43]

The Donora, Pennsylvania, smog disaster was a turning point in American environmental history. A temperature inversion in the western Pennsylvania community trapped nitrogen dioxide, sulfuric acid, and fluorine being emitted from two U.S. Steel facilities. The deaths of twenty people were attributed to the event. Air pollution, previously considered a nuisance at worst, and an inevitable cost of industrial development, was now viewed as having potentially fatal consequences. The 1948 event would reverberate throughout the nation and become a symbolic representation of the need for environmental protection, and Allied's managers seemed to be getting the message that they would at least have to appear to care about public health. The company promised that it would limit its dust discharges by the end of the year with the installation of an abatement system to capture both fly ash and lime dust. The *Syracuse Herald-American* described this as "good news."[44] The paper followed with two very positive editorials, congratulating both the company for its decision to spend $500,000 on the required equipment and the citizens of the community for working actively to mitigate smoke abatement in the community.[45] The case seems unusual in that a highly polluting company responded to community concerns proactively, without any legal action being initiated by state or local authorities. But such optimism turned out to be premature.

Meanwhile, the water pollution problem had still not been addressed in a significant way. At the end of the year, the City Planning Commission proposed a new municipal authority to deal with the lake's pollution problems, part of which would be to experiment with "shrubbery" that would hide the most evident solid waste problems.[46]

Allied Responses and Town of Camillus Resistance

In early 1949, the Solvay Company, which had been a division of Allied Chemical and Dye Corporation, announced that it would stop releasing effluents directly into Onondaga Lake. The company promised to spend $1 million, not only to redirect its outflows away from the lake, but also remove the unsightly pipes that facilitated the process. Director of Operations Carlton Bates said that the company was making the change "to cooperate with county and state administrative authorities interested in improving the lake shore area." The waste would now be carried by pipe two miles away from the lakefront and deposited in the field near a railroad yard not far from Nine Mile Creek and Geddes Brook. The pipe would cross the border into the town of Camillus, a growing suburban area, known for its forested hills, a place popular among area hikers and bird-watchers.

As for the beds of stored waste that still populated the shoreline, the company reassured residents, "After a time rain and snow water will leach out the soluble materials, such as salt, and natural vegetation may be expected to grow on the bed. This has happened to all the older Solvay wastebeds, which are located in the vicinity of the plant and further south along the lake shore."[47] The beds, which had been in active use from 1915 until 1944, had created a plateau that rose sixty feet above the lake and was composed primarily of discarded limestone, salt, and calcium compounds. The substances, while not in-and-of-themselves toxic, were infused, to some indeterminate degree, with toxic chemicals such as benzene and toluene in addition to heavy metals, including mercury.[48] Both local newspapers, the *Syracuse Post-Standard* and the *Herald Journal*, offered tentative support for the plan.[49]

Solvay's move to redirect its waste away from the lake created a conundrum for the city of Syracuse, which had been mixing its sewage sludge in with Solvay's industrial waste. The Solvay wastes were believed to kill the bacteria that thrived in the sewage, keeping the city in compliance with the state's public health law. The city would now presumably have to find another mechanism for disposing of this material or of killing the bacteria.[50]

By June 1949, the company's cooperative tone changed markedly as Allied began to threaten that it would move the plant if it could not get the waste issue resolved. Residents of the town of Camillus were balking at the plan to pipe wastes into an abandoned farm within the town's borders, claiming that zoning restrictions prevented it. The company was displeased that its planned multimillion dollar upgrade of the processing facility was being impeded by the opposition. A West Side Citizens Association was formed and held a hearing at which company spokesperson Carlton Bates testified that Solvay was considering relocation. After the hearing, the committee issued a statement suggesting that it was "astounded to learn the immanence of this economic danger to Onondaga County, and even more astounded that the situation is so little understood by the business and laboring interests of the city."[51] But Walter Welch, a Republican and member of the Onondaga Lake Reclamation Association, challenged the Citizens Committee, accusing it of being a front for the company and questioning its findings. The conflict between Welch and the committee was characterized as a "showdown" as he "hurled down the gauntlet" at the citizens' association.[52] Letters to the editor of local newspapers indicated that deep divisions were forming within area communities."[53]

The Citizens Committee responded through the press that it had been involved with a series of local issues and was not simply a mouthpiece for Solvay Process. It criticized Welch for using language "not commonly used by gentlemen in discussing public questions."[54] Welch's exact words were not reported, so it is unclear whether the Committee was objecting to the general tenor or his statements or whether he used profanity. Whatever the case, the incident indicated growing community antagonisms over Solvay practices.

William Maloney also became involved in what was becoming a very heated public dispute. Maloney was adamant that "Onondaga Lake is going to be reclaimed, all opposition to the contrary." Maloney argued that the lake, which could be a community asset, posed a serious public health problem in its current condition. "Onondaga Lake," he stated, "is a natural body of water created by God in his Plan of things, and no intelligent person would think that it is intended that the lake should ever be ruined by so-called civilization." Moreover, he regarded it as an "obligation" that the company dispose of its waste "without infringing on the rights of others." He accused the company of "hiding behind an employment of persons" in the failure "to curtail its despoilment of the lake."[55] Walter Welch also responded in a letter to the editor defending himself against the "unfounded charges and insinuations" made by the West Side Citizens Association."[56] Soon thereafter a "disgusted Syracusan" asked in a letter to the *Post-Standard*, "Does Syracuse want its people employed or do they want the lake cleaned up?"[57] That seemed to be the central question at the heart of the dispute. For the first time, the choice seemed to have been made abundantly clear: jobs and economic security versus protection of the area's natural environment, a conflict still often seen as at the heart of many of environmental conflicts in the twentieth century.

In October, and in response to the increasingly heated controversy, Solvay Process took out an advertisement in the *Post-Standard* asserting that it might be required to move the plant were it not given access to the zoned property in Camillus.[58] Walter Welch responded via a letter to the editor suggesting that the proposal to move the State Fairgrounds may have been the result of collusion by Solvay Process and some state officials, since that proposal was considered by some to be a means of turning the fairgrounds into a waste dump.[59] That November, in a statement before the Camillus Town Planning Board, the company seemed to take an increasingly hard line. It noted not only the potential loss of jobs in the area (including residents of Camillus itself), but also the tax receipts generated by the plant, along with a loss of employment by the railroads and the Portland Cement Company. It noted the "interdependent" nature of the economy in the area, and the picture painted was one of devastating economic loss.[60]

Welch responded, congratulating the newly formed Citizens' Council of Camillus (not to be confused with the pro-Solvay Citizens' *Committee*) for its stand challenging the company's demand for a change in the zoning ordinance. Welch also had discovered two deeds in the county courthouse. The first deeded a parcel of land from the old Erie Canal, transferring it to the Village of Solvay for more than $38,000. Another deed, executed the same day, two minutes later, transferred the same parcel of land from the Village to Solvay Process for $1. The deeds, he asserted, "disclosed that land which would and should have provided a five-minute boulevard from the city to the fairgrounds, and a perfect truck route to the west, was traded away by our elected representatives for a pittance."[61]

On December 1, the Town of Camillus Planning Board recommended a change in zoning for the parcel of land owned by Solvay. The Board report cited public support for the change and the economic interests of the area. Two board members dissented, contending that alternative sites had not been considered and that this would render the land forever useless for any other activity.[62]

On December 8, the State Water Pollution Control Board met at the Onondaga County Courthouse to begin hearings on the lake's water quality. The state was again attempting to determine what uses might be possible for Onondaga Lake water. Drinking had been ruled out, but fishing was still on the table, even though fish caught from the lake were "oily." Various industrial uses were also possible.[63]

Somewhat surprisingly, when the Camillus Town Board met, it rejected the Planning Board's recommendation and voted not to change the town's zoning ordinance. The Camillus Citizens' Council congratulated the board for its actions, stating that, "More and more thoughtful people are beginning to realize that the wanton destruction which has characterized the actions of the company for the past 67 years is not necessary and need not be accepted."[64] Solvay Process did not passively accept the town board's decision. Rather, it released a report under the auspices the Allied-friendly Citizens' Committee, making a series of recommendations and threats. The Syracuse Common Council, at a point when the City of Syracuse had control over the land, had left open the option of selling the land to the company by deleting the date on an earlier offer, resulting in an open-ended opportunity for the company to purchase the land. Given this, the Citizens Committee, recommended compromises to make the sale more palatable. Among these was to set up a park, complete with a swimming pool that would act as a buffer between the waste disposal facility and the rest of the town. It also offered to develop "conservation areas" around the waste site. In turn, Solvay Process would work to keep the waste within the boundaries of the disposal site and that it would stop sending its pollutants into Onondaga Lake. If, on the other hand, the town did not agree to the sale, "the county of Onondaga and its residents will suffer in every phase of community life with direct economic loss of disastrous proportions."[65] The Camillus Citizens Council labeled the report a "sham."[66]

At a public hearing held January 19, the dispute became more complex and the interests more fragmented. Some residents in the town of Dewitt, to the south and east of Syracuse, wanted to invite Solvay Process to move all of its operations there. Dewitt had much to lose if the nearby Jamesville Quarry was closed, but its offer seemed impractical. For one thing, tiny Pools Brook would not likely have the capacity to accommodate Solvay waste disposal.

Walter Pope, a spokesman for the company, said that if the land in Camillus was not available it would seek to acquire tracts of land in the Town of Geddes, which runs north and south along the lake and constitutes a larger area than Camillus (surrounding the Village of Solvay).[67]

Meanwhile, the indefatigable William Maloney continued his public conflict with Solvay Process through letters to the editor, at one point accusing Allied Chemical of being a foreign corporation whose loyalty to the U.S. was suspect.[68]

The Town of Geddes Joins the Resistance

While the dispute regarding what to do with Solvay wastes remained unresolved, the City of Syracuse was attempting to get a handle on the sewage that was now "pouring into" Onondaga Lake. (Approximately 20% of the lake's inflow was being funneled through the inadequate sewage treatment plant.) In June 1950, when Solvay workers went on strike, the city was unable to mix its sewage with waste products from the plant. The city's response was to inject 5,000 pounds of chlorine into the waste each day, in an apparent attempt to disinfect it. The $130 cost was significantly more than the usual $30 allotment.[69] At one point, the city had to negotiate with striking workers to allow chlorine, being held on the company's property, to be released.[70] The State Board of Health approved dumping the sewage into the lake with the addition of chlorine. Residents in the Town of Geddes objected, but the state asserted the dumping could continue for at least a month without endangering anyone's health.[71] However, the conditions were so obviously foul that even the Liverpool P.T.A. took a stand, asking the city to address the issue.[72] Quite simply, the smell of the lake was overwhelming. The impacts of the strike pointed to a larger problem looming in the future. If Solvay were allowed access to a landfill for its wastes, then it would close the wastebeds near the lake, and the city would need some alternative means of "sterilizing" the raw sewage flowing from its treatment facility. The city's decision, decades earlier, to install a Dorr clarifier as a minimal form of wastewater treatment largely in response to cost factors, was now coming back to haunt it in a very obvious and significant way. At this point, given the mixing of Solvay wastes with county sewage, any town that agreed to accept the former would implicitly (perhaps explicitly) be required to accept the latter. In other words, a town would open itself to tons of contaminated calcium chloride and tons of untreated, or lightly treated, Onondaga County sewage.

The crisis of where to dispose of the sewage seemed to be enough to spur the county to make steps toward resolving the sewage problem, however. In July, the Onondaga Public Works Commission announced that it would propose a new sewage system for the southern end of the lake that would eventually eliminate the dumping of raw sewage.[73] This began a decades-long saga of trying to find a more effective way of dealing with the county's sewage.

The State Pollution Board continued to hold hearings and make inspections in its quest to categorize the lake according to one of seven categories, from least-polluted (for drinking purposes) to most-polluted (only industrial purposes).[74] A preliminary finding by the state superintendent of sewage found that oxygen levels in the lake were so low that the only fish it could support were carp. The Board has discovered the obvious: the lake was at the bottom of any safe water usage scale.[75]

In August, the striking workers reached an agreement with the company, and the city hoped to return to the practice of mixing its sewage with Solvay waste.[76] In October, it resumed the mixing/dumping process.[77] Albert Pope, an advocate of moving the Solvay waste to Camillus, objected to the renewed dumping as a member of Lakeland, a community within the town of Geddes, and the area that would be most impacted by the resumption of the process.[78] The two towns now seemed to be at odds with one another as to who would or should accept the Solvay wastes. The Town of Geddes passed a resolution objecting to the renewed dumping, and Pope stated that he would personally fill out a warrant to require the city to stop. But the city had state authorization. In fact, it was under orders from the state to resume the former process, and thus could effectively ignore Pope's warning, which it did.[79] *The Post-Standard*, in an editorial, showed little sympathy for the town, suggesting that the mixing of sewage with limestone waste to form the wastebeds was a reasonable solution to a difficult problem.[80] At the same time, the Syracuse Common Council supported plans to move forward with the completion of a sewage treatment system,[81] but the town of Geddes was holding firm, and the city honored its decision to prevent the dumping of Syracuse city sludge within its borders. As a result, the city continued dumping raw sewage into the lake.

Nelson Pitts, a city engineer who was responsible for completing yet another survey of the lake, gave his preliminary appraisal in July 1951.

He stated that the lake's condition had deteriorated markedly since the city of Syracuse began dumping raw sewage into the lake a year previously. On the northern part of the lake, the water was "milky" with the distinctive odor of Solvay sludge. On the southern end, "the lake gave evidence of varying degrees of pollution," and "an unmistakable odor of sewage." Pitts concluded Onondaga Lake's condition was a "disgrace to the community."[82] The Onondaga Lake Purification Commission, set up previously to study the matter, simply recommended that the county public works commission and the city work together to build a sewage treatment facility that would resolve the problem, at least with regard to the city's sewage. But the city seemed unwilling to commit the necessary funds, drawing the ire of activist William Mahoney, who had withdrawn from his life as an environmental activist for nearly six years, but who was drawn back into the fray over the sewage issue.[83]

By 1952, Nelson Pitts had become a key public figure pushing for the lake cleanup. He publicly criticized Syracuse's intransigence in committing to a sewage disposal plant, and he contended that the community could have a lake as clean as they were willing to pay for.[84] But, given intransigence in the town of Camillus, city officials were pushing the town of Geddes to give at least temporary permission to dump the sewage in Solvay wastebeds that were located there.[85] But gas station proprietor Walter Pope, whose "almost single-handed" effort to stop the extension of a pipe that would have allowed the dumping, had put fear into the heart of the Geddes town board. Solvay Process could not, he asserted, violate the town board's finding in the case. Pope seemed to be negotiating with the city of Syracuse about a temporary agreement to dump, but only with the understanding that a new sewage treatment plant would be completed.[86] Moreover, the city apparently did not want to antagonize the Geddes town board, with which it hoped to cooperate on the building of a joint sewage treatment plant. Still, the town was under considerable pressure to reach some kind of at least temporary agreement with the city, because, as attorney Sidney B. Coulter noted, members of the board felt a "moral responsibility" for the continued dumping of sewage into the lake by the city.[87]

Soon the city and the town came to an agreement for extending the sewage pipe to the new Solvay wastebeds in the town of Geddes, and the issue seemed to have been at least temporarily resolved.[88] By April the pipe extension, under the direction of engineer Pitts, was completed, in time for an upcoming sailing regatta scheduled to be held on the lake. Once completed, the city could return to its old method of dumping its sewage in with the Solvay Process waste, which was said to "neutralize" it in "a matter of about 90 seconds."[89] Asked if a sewage treatment plant could be built that would eliminate the pollution problem in the lake, Pitts stated, "Yes, it should have been done ten years ago."[90]

In May, the State Water Pollution Control Board released its findings, and much to the chagrin of city engineer Pitts, divided the lake into four different parts. The upper end of the lake would be considered okay for "bathing" and "recreation" (a "B" designation); the middle part for fishing, but not swimming (a "C" designation); and the lower part, only for industrial purposes, but not fishing or swimming (a "D" designation). Pitts's response was, "I know only about three kinds of water—hot water, cold water and dirty water. Put together, they're still dirty water."[91] In retrospective, the designations for the upper and middle parts of the lake seems wildly optimistic.

Also in May, in spite of the completion of the pipeline to Geddes, the town was dragging its feet, with the town board unwilling to make a commitment to accept the city's sewage. On May 14, the town board convened, and decided against rescinding the ordinance to prevent the dumping of city sewage. Consensus was such that no vote was even taken. Apparently, part of the reason for the retraction was the state's classification of the part of the lake near Geddes as class B, thus not allowing swimming, which town residents were hoping to have permitted. Moreover, the town's attorney expressed little faith in promises made by the city to build a sewage disposal facility to ultimately resolve the problem.[92] The tone of the meeting was described as "tense" and "bitter," as residents of Geddes, including Walter Pope, railed against the city.[93] The Town of Geddes had joined the Town of Camillus in fending off efforts to use property within its boundaries as expanded waste disposal sites.

While this was occurring, Allied was dedicating a new state of the art, multi-million-dollar soda ash facility, making a move out of the area extremely unlikely.[94] And the county was moving ahead with plans to improve the park area that surrounded the lake.[95] Area sports enthusiasts also began to publicly state their support for cleaning up the lake. Over time, they would be a key constituency pushing for lake cleanup.[96]

In a hearing before the State Pollution Control Board, held at the end of June, Redmond Keyes, a representative of the Onondaga Federation of Sportsmen's Clubs, presented a petition signed by 3,000 members demanding that the lake be brought up to class B levels, so that fishing in the lake could resumed.

At the same meeting a member of the Syracuse crew team also spoke up in favor of restoring the lake, as did Mrs. Henry C. Fadden, President of the Liverpool Community Council, who described the lakefront as "beautiful but dangerous." Engineer Pitts, according to the article, "again found himself at odds with the Corcoran administration and Common Council" as he demanded that Syracuse and the surrounding communities "refrain from using Onondaga Lake as a cesspool."[97]

The Eternal Search for a Sewage Disposal Solution

In October 1952, Mayor Tom Corcoran, apparently because negotiations with the Onondaga County Planning Commission were going nowhere, decided that the city would go it alone. The city, he said in a letter to all members of the Common Council (eight Republicans and one Democrat), would go forward with the construction of a new sewage disposal plant. The job of planning and constructing the plant would be given to Nelson F. Pitts, with an estimated cost of between ten and twenty million.[98] Frank Early predicted, or at least hoped for, a new "Golden Age" for the lake, returning it to its former glory as a recreation and amusement destination in the 1890s.[99] In October, Pitts was praised for his work on smoke abatement in the city by the Chamber of Commerce, as part of "National Cleaner Air Week."[100] The mayor jumped on board by declaring a local "Cleaner Air Week."[101]

Mayor Corcoran's decision to have the city take the lead on the sewage disposal system met with some immediate problems. First, there was a city ordinance that gave the county control over the city's sewage disposal. Second, if the city and county split the costs, the city would pay more, because it would have to purchase land inside the city limits, making it more expensive than if land were purchased in one of the surrounding towns.

Directed by engineer Pitts, work continued to go forward to extend the pipeline that would bring Syracuse sewage to the newer Solvay wastebeds.

But Pitts and the city were on a collision course with the town of Geddes, which was waiting until the town boundaries were actually crossed until it took legal action against the city. The town was also toughening up on Solvay Process, preventing the dumping of waste with 2000 feet of public buildings, dairy farms, or private residences.[102] Mayor Corcoran, trying to take some of the political heat off of his administration, responded to these controversies by firing Pitts, who had served as city engineer for 27 years. This strategy failed, however, and the pollution controversy contributed to Mayor Corcoran's loss in the November election.

Succeeding Mayor Corcoran was newly elected Mayor Don Mead. When he and the new Republican administration came into office, they developed a plan to divert the city's sewage into trenches dug into older Solvay wastebeds, running along Hiawatha Boulevard. Previously, city sewage had been mixed with freshly disposed Solvay waste, so the new proposal was something of an innovation. The company gave the city permission to conduct this "experiment," which would end the dumping of raw sewage into the lake. Democratic members of common council, in response, organized a "clothespin campaign," asking local residents in the Second Ward, to send letters of protest to the mayor's office containing a clothespin to represent the stench that would be created by the series of trenches that would be dug into a twenty-three-acre area along the lakeshore.[103] Undeterred, the city went forward with the plan, and Mead, after inspecting the area a week after its initial completion, declared it to be working well. According to the *Herald-Journal*, "The mayor said the old Solvay Process Company wastebeds appears to be acting as a perfect filter to separate solids from liquid, even working in a lateral direction as the liquid seeps through the porous soda ash of the wastebeds."[104]

At the same time, the State Pollution Control Board released its recommendation that the lake be cleaned up so that 62.5% of it be suitable for swimming, with the lower (southern) third of the lake, the area in which raw sewage was then being dumped, required to move toward a C rating, which was suitable for fishing. The board released a map with the recommendations, featuring a series of lines to mark precisely which areas would be recommended for which activities.[105] The state then demanded that the city take action on studies that addressed how to rectify the situation by June 1, 1954. Mayor Mead publicly acceded to the state's request.[106]

After receiving proposals for several sewage disposal systems, Mayor Mead recommended adoption of a metropolitan sewage system that would consolidate town sewage systems into one plant on Onondaga Lake, treat it to remove the sludge, which would then be digested and incinerated, with the remaining liquids being dumped into "the nearest practicable receiving waters, exclusive of Onondaga Lake." The only other option was the Seneca River into which the lake empties, but which is several miles from the city's borders.

The total cost to the city would be a little over $15 million. The proposed system would have the advantage of ending the disposal of Syracuse sewage sludge into the Solvay wastebeds, but it would only involve primary sewage treatment.[107] In other words, it would mean constructing an updated version of an Imhoff tank, which involves settling of solids alone, with no anerobic bacterial action to digest it. *Post-Standard* columnist Grace Lewis seemed to be one of the few with a memory of how the whole mess went down. "The fact is," she stated, "the city fathers snuggled with the Solvay Process (by then AC & D) brains in 1911 or so, and hit upon a scheme. It fitted nicely, providing the public kept on its usual glasses, and history has proved sadly that the public was well-equipped." Lewis questioned the "mystical" neutralizing properties of the Solvay wastebeds on Syracuse sewage. Allied Chemical and Dye, she noted, "has never refuted the claim of 'neutralizing' by chemicals and all the time it has talked of 'inert substances' in its wastes."[108] In an editorial, the *Herald Journal* wondered whether communities along the Seneca river would accept the outflow of sewage from which only 35% of the contaminants had been removed.[109]

Various objections were raised to the new sewage plant, partly based upon the equitability of financing between the various entities in the county. But at one meeting, engineer Pitts, by then retired, returned to raise objections to the proposed system being only a primary disposal system, leaving large quantities of sludge to be disposed of. Pitts accused the city of returning to a system that would inevitably result in the sludge being deposited in the Solvay wastebeds, a claim that was denied by Earl F. Obrien, a consulting engineer for the firm proposing the plant. Obrien suggested that the sludge was rich with "protein" and other materials, and it could be used as fertilizer for growing vegetation on the west side of the lake where the wastebeds had been expanding for decades, if it were turned into a park.[110]

By May 1955, however, New York State seemed to take an increasingly tough line. The Water Pollution Control Board listed thirty-four entities contributing to the problem in its "Comprehensive Plan for the Abatement of Pollution in the Waters of the Onondaga Lake Drainage Bed." Among these were the City of Syracuse, various villages near the lake, a variety of private industrial concerns, including Solvay Process and Carrier Corporation, and a group of small businesses such as the Onondaga Furniture Fixture Company and the James Rock Spring Restaurant. The state asked for voluntary compliance, but threatened legal action if appropriate plans for mitigation were not put into place.[111] But by later that summer, when a hearing was held by the board, Colonel A. J. Dappert, its executive secretary, expressed that they were "delighted" with the Public Works Commission's progress in creating a West Side sanitary district, and upgrading and expanding existing facilities.[112] Solvay Process was congratulated for its decision to give the state a gift of the property upon which the wastebeds were located to build a new parking lot for the State Fairgrounds.[113] The state seemed to be backing away from its earlier threats, expressing its satisfaction with efforts that, in truth, offered little of substance. Still, the movement forward on a new sanitation district gave cover to state authorities, and things seemed to quiet down. The crisis that was threatening to engulf Solvay and other industrial operations was averted, at least temporarily. Nevertheless, significant problems with Solvay Process's pollution had not been addressed in a meaningful way.

In August 1957, state authority reentered the discussions when the State Board of Health asked a number of industrial firms to stop dumping pollutants into area lakes and streams. Among those cited was Solvay Process, due to its continued dumping into Onondaga Lake as well as Nine Mile Creek, the main tributary feeder into the lake.[114] The Health Department also began to take a closer look at Oneida Lake, which had previously been neglected with all the attention focused on Onondaga.[115] The overloading of the sewage treatment plant in Liverpool became an increasing concern of the State Health Department. Despite these concerns, Village officials did not even bother to show up for talks with the State Health Department on the matter, essentially ignoring its concerns.[116]

Geographies of Contamination

The intertwined contamination problems in Onondaga Lake in the middle part of the twentieth century are unique. Nowhere else is there a documented case where public authorities worked with a private corporation to encourage it to continue its disposal practices because they hid the malodorous impacts of an antiquated and inadequate sewage disposal system. The claims of company officials that their calcium chloride piles actually cleaned the lake were, of course, greatly overstated, if not completely absurd. That is clear to any informed observer today and should have been to those involved during the period in which this occurred.

Residents of growing suburbs were aware of the nastiness associated with these combined disposal substances, and they wanted nothing to do with them. Camillus was a town identified with parks and forested hills, and even though the Solvay wastebeds would be expanded just across its border, residents successfully organized to keep them out. Their relative affluence, access to their own elected officials, insulation from the economic power of Allied, and social capital were crucial in the success of their effort. Public officials in the town of Geddes, which included Solvay and other industrial operations in its borders, were more ambivalent. The construction of a pipe farther north along the town's borders was agreed to by the county board. But the board eventually retracted its decision to allow waste to be deposited close to the growing affluent suburban village of Lakeland.

The leaders of the nascent Onondaga County local environmental movement that resisted Allied's proposals represented a diverse array of interests: William Maloney was a banker, Walter Welch a landscape architect, Walter Pope a gas station proprietor. This was the era before environmental nonprofits emerged to confront urban pollution problems. It provides evidence of the deep roots of American environmentalism, which developed in complex ways to address local contamination issues in the decades before the appearance of environmental movement writ large in the 1960s. Moreover, these activists had an impact, influencing both county and state officials. And, while they may not have been successful in their goal of cleaning up Onondaga Lake, they provided the foundation for the environmental activism, scientific investigations, and legal and policy decisions that would follow.

Had the only effluent involved been Solvay wastes, which Allied claimed were inert, one or other of the targeted towns might have agreed, but accepting sewage to be "cleansed" by the Solvay wastes was too much to bear. In the end, it was the lake itself that suffered, because the county had no alternative but to dump raw sewage directly into it, further antagonizing residents all along the lakefront to Solvay disposal practices. Ironically, because Allied's facilities were such that they provided a border between the Village of Solvay and Onondaga Lake, the Village was shielded from the expanding waste piles, and thus those with the largest economic stake in the outcome were not direct parties to the decision-making.

4

Ecologies of Contamination

A Huge, Stinking Mess and a Wave of Opposition

In September 1966, the body of Ellger Tomonia was found in Onondaga Lake, the victim of an apparent drowning. Tomonia, who had been living at the Union Rescue Mission, had left a sport coat, his shoes, and three handkerchiefs on shore, while he went swimming in his khaki pants and plaid shirt. Only a tube of toothpaste, a razor, and two razor blades were found in his pockets. The exact circumstances of Tomonia's death were never determined. As serious as the pollution problems were in the lake, they would not have been severe enough to kill a person, yet it is not a large leap to see his death as a symbol of the lake itself. The lake was moribund, under mortal threat, the result of years of unregulated dumping of chemicals and sewage, a toxic mix that was finally beginning to be seen as an unacceptable stain on the reputation of Onondaga County and the City of Syracuse.

As noted previously, the 1940s and 50s marked a period of important, if often unrecognized, environmental organizing. Often this took place, as in Onondaga County, in local communities, as citizens from different social and economic backgrounds, grappled with the very immediate ecological problems that they faced. The 1960s and 70s were a period of more widespread environmental awakening. While the groundwork to Rachel Carson's work had been laid by multiple other scientists and activists that preceded her, with the publication of *Silent Spring*[1] in 1962 a public debate erupted over the use of chemicals, especially DDT, that had in the post-war period come into extensive use. President John F. Kennedy entered into the debate when he publicly endorsed Carson's findings. The late sixties and early seventies saw an explosion of environmental legislation: the Clean Air Act (1963), the National Environmental Policy Act (NEPA, 1969), the Clean Water Act (1972), and the Endangered Species Act (1973).

It is within this context that responses to the contamination of Onondaga Lake should be examined. The macro-level changes in the political climate influenced how the area's pollution problems were being evaluated and addressed. At the same time, attitudes and actions specifically related to the lake crystallized conflicts, some of them intractable, that would impact broader regional and national debates. The politics of Onondaga Lake were being defined by an emerging social movement, while simultaneously contributing to the shape that movement would take.

The period of the 1960s and 70s represented an environmental awakening, but that awakening held an element of naivete, with activists often underestimating the difficulties of undoing the damage caused by decades of industrial production. Over time, as the extent of Onondaga Lake's contamination problems became more evident, naivete shifted to a more mature recognition that any lake cleanup would be a difficult and expensive undertaking. It would take several decades for this realization to fully take hold and for realistic plans to be developed and implemented. And the adequacy of the final outcome remains in dispute, as we shall see.

A Web of Denial and Collusion

Like all lakes, Onondaga Lake is part of a larger watershed. Multiple tributaries drain into it. At the southern end of the lake, Onondaga Creek, which runs through the city of Syracuse, provides most of the lake's inflow. On the southwest is Harbor Creek. Nine Mile Creek runs through the village of Solvay and became a convenient receptacle for wastes from Allied, Crucible Steel, and other industrial concerns. East Flume is a channel that runs adjacent to one of the main waste-beds and became the primary route by which Allied discharged its effluents directly into the lake. Tributary 5A, also a channel, carries water from a natural spring around the Semet residue ponds and into the lake. The ponds were originally used to dispose of Allied waste but became a repository for tar from a benzol production plant that began operation in 1917, creating an amalgam of toxic goo. The ponds are surrounded by a dike, into which was funneled Solvay wastes, sometimes mixed with sewage sludge from the nearby treatment plant. It was constructed from found materials, including "concrete rubble, old electrolytic cell parts, ashes, cinders, soil, Solvay Waste, bricks, and stone."[2]

On the northern side of the lake are two creeks, Sawmill Creek and Bloody Brook, which being ten miles from Solvay, had remained relatively free of industrial toxins. Entering the lake from the southeast is Ley Creek, into which was discharged sewage and industrial effluent from a General Motors plant. Buttermilk Creek flows into Lake Oneida, but it passes near the quarry that provided the limestone essential to soda ash production. Runoff from the quarry found a path to relatively pristine Oneida via this route. The complex and rich ecological system represented by the Onondaga watershed made multiple avenues available for disposing of chemical contaminants and other waste products, expanding and complicating eventual attempts at a cleanup. Expanding suburban developments that sprang up in the area in the 1960s and 1970s brought increasing numbers of residents who were aware of the reach of the contamination and the potential harms that Allied and other industrial operations posed. These provided the ground for further political resistance that had been initiated by residents of nearby towns of Camillus and Geddes in the previous decades.

Given the extent of waste disposal within a confined watershed and in proximity to a densely populated urban area, it was only a matter of time before area residents would demand some kind of action. But lacking any comprehensive plan, and given the significance of Allied and other industrial facilities to the economic life of the area, residents' efforts at redress were scattered and ineffective. The 1960s through the 1980s represented a period when environmental harm was recognized not only in tangible ways by the communities impacted, but also in more systematic ways via scientific investigation. Various constituencies emerged to demand action, but initially, industry attempted to minimize or ignore them. As those strategies proved ineffective, companies launched public relations campaigns to present themselves as good corporate citizens. Meanwhile, over time, regulatory authority at the state and federal levels began to provide mechanisms to support activists and other interested parties, with varying success. The coalescence of all these strands—recognition by the state of contamination problems, resistance on the part of affected communities, and regulatory responses—marked the appearance of an environmental consciousness that would gain influence as it expanded.

By the end of 1958, the city and county appeared to be moving forward with a West Side sewage district project. Authorities issued the permit for the building of a digester to deal with the sludge that was being transported over to the Solvay wastebeds. The Public Works Commission solicited bids on sewage flow units, a major part of the project.[3] A new pumping station was also opened that would send effluent from the Village of Solvay and Solvay Process into the plant that had been owned by the city of Syracuse, but was now under the control of Onondaga County. This would reverse the process which previously had sent sewage sludge to the Solvay wastebeds.[4] The rationale for the reversal was unclear, since Solvay wastes were primarily calcium chloride (infused with a variety of toxins), a solid that would not break down via sewage treatment. Authorities continued operating under the belief that Solvay wastes would "neutralize" the area's sewage. Even as the project went forward, recognition that it wouldn't provide immediate relief to the lake was emerging. One county official predicted that fish in Onondaga Lake wouldn't be fit for consumption "by 1980."[5] Ultimately, the county was caught in a web of denial and collusion that had long marked its relationship to Allied Chemical.

The sewage facility itself was only part of the problem. A $47,000 grant awarded to the city in 1960 focused on the inadequacy of the broader sewer systems. The study found that fully 40% of the sewage escaped from the sewers and never made it to treatment, instead flowing directly into Onondaga and Harbor Creeks.[6] The town of Camillus experienced problems with its pumping station, which was allowing raw sewage to flow into Nine Mile Creek.[7] Onondaga County had been experiencing an almost complete breakdown of its entire sewage system, the effects of which pervaded the Onondaga Lake watershed.

Sports fishermen began to emerge as a key constituency calling for a cleanup of the regional watershed. Butternut Creek, into which a variety of industrial concerns dumped their waste, flowed past the Jamesville quarry, operated by Solvay Process.[8] At issue was the fact that the creeks were interconnected to a variety of other contamination problems, so addressing them as individual entities would be inadequate.

Air pollution also continued to be a significant problem. Residents of Syracuse and the village of Solvay complained of smoke emissions from Allied and Crucible Steel. Air pollution, given its diffusion over multiple affected communities, generated widespread public opposition. Officials in the city of Syracuse threw up their hands, asserting that they had no power to control Solvay's air pollution emissions, given that it was beyond the city's jurisdiction.[9] Solvay responded by sending engineers into the community to discuss how it was addressing the issue, but residents detected little improvement.[10] The federal government had taken note of air pollution problems as early as 1955, with the passage of the Air Pollution Control Act, which recognized potential health impacts of it, but confined the government's role to providing information. An expansion of the act in 1967 provided more funding for research and authorized states to develop their own standards for air pollution abatement, but not one state took the offer. Before passage of the Clean Air Act of 1970, states and local communities had no real power, and little alternative but to rely upon the good will of the polluting industries.[11]

A Shift in Political Winds

In the mid-sixties, elected public officials began to recognize that responding to local environmental concerns might be politically advantageous. Key political figures, recognizing an opportunity to gain support, adopted lake cleanup as a priority. Again, this followed national trends.

In January 1966, James McCarthy, a Democratic member of the Syracuse Common Council (16th ward) proposed a committee to study the sources of lake pollution and mitigation strategies. The committee would be comprised of "technical experts" from Syracuse University private industry, including Allied Chemical itself.[12] Area sports fishermen, without formal technical training, but often with an intimate knowledge of lake conditions, continued to lobby for the lake's cleanup as well.[13] A coalition of interests was emerging, often from different class and educational backgrounds, the focus of which was understanding and mitigating the lake's multiple pollution problems.

Allied seemed to be aware of potential threats to its operations. In February, at a hearing held by the State Water Resources Commission on the water quality of streams that flowed into Onondaga Lake, the company pushed back. The commission was attempting to determine if the status of the creeks, currently rated "F," and thus not recommended for any use whatsoever, could be upgraded to a "D," to make them available for farming and industrial purposes. Regional State Engineer Howard Gates did not propose any mitigation measures to alter the quality in the creeks, so the rationale for the change was questionable. Still, he did point to Crucible Steel and Solvay Process as major offenders for contaminating the creeks by dumping acids, alkalies, and other chemicals into drainage ditches that then leaked into the area's watershed.

Crucible didn't bother to send a representative to the hearing, but Solvay's Production Director, W. Y. Lobdell, contended that any upgrades were unrealistic and that the dumping by Solvay Process was actually beneficial to the lake. He asserted that "nothing coming into the streams from our plants is harmful to health [and] that much of what comes in has substantially beneficial effects under present conditions in the waters." Lobdell grounded his claim in the hypothesis that Solvay's sludge piles were helping to filter the area's sewage, which the county was unable to properly treat. He had a point about the county's responsibility, but the assertion that Solvay wastes were actually good for the lake, a tactic that had previously been deployed by Allied, was losing its potency, and it did not mollify lake cleanup advocates.

While County Sanitation Commissioner Uhl T. Mann did not seem to object to reclassifying Onondaga tributaries to a D category, he noted the obvious: Changing the classification did nothing to change water quality. He thought it impractical for the state to upgrade any of the eight streams feeding Onondaga to anything above a D without improvements in water quality. In his words, "Classifying water 'B' (for bathing) doesn't make it 'B' water."[14]

John Mulroy, Onondaga County's first county executive, assumed the office in 1962, after a new charter created the office that he would hold for more than twenty-five years. Mulroy was destined to become a powerful figure in local politics. He would be the force behind many improvements in the county: educational, cultural, and environmental. But his career would eventually be tarnished by corruption charges in the late seventies and early eighties.[15] Nevertheless, he became a central player in Onondaga Lake cleanup efforts, albeit with a mixed record.

Soon after assuming office, Mulroy jumped on the bandwagon, with those who wanted to determine the extent of lake pollution, by announcing the formation of a new technical committee.[16] The Onondaga Lake Scientific Council's purpose would be to research the problem and provide advisory information to the county based upon its findings. It was an important step forward to have a group whose primary mission was to study the lake. Lake advocates, such as members of the Onondaga Lake Clean Water Association, were supportive.[17] However, while Mulroy showed an initial interest in investigating and addressing lake contamination problems, there were limits to his advocacy.

When the Onondaga Lake Scientific Council proposed a twenty-three-point plan involving the expenditure of $25.5 million dollars for lake cleanup, in retrospect a fairly modest sum, Mulroy rejected it as too expensive, and announced a nine-point alternative proposal, to which no dollar figure was attached.[18] He also agreed to allow Syracuse University to dump 100 tons of ash from its steam generating plant onto the Onondaga's shoreline. Deputy County Executive Harry W. Honan assured critics that the ash would not cause further pollution problems to the lake, and that oil (!) would be put on top to keep it from blowing around. Daniel Jackson, head of the council and recently appointed by Mulroy, questioned the wisdom of the decision, and noted that the council had not even been consulted. "If the services of the experts on water on the . . . council are not to be sought or used, why have this council?"[19]

About this time, it slowly began dawning on Allied executives that, given the increased scrutiny, their corporate image was and would continue to be tarnished by the lake's pollution. Claiming that Solvay Process wastes were beneficial to the local ecology would probably not fly for much longer. The result was a public relations campaign, in what could best be described as an early attempt at greenwashing. The company created a speakers' bureau, housed at a facility in Syracuse, that was available to give talks on the Solvay production process, including its air and water pollution abatement programs. The company claimed that 10,000 people, including many students, had taken advantage of the service.[20]

By the mid-1960s, following larger national trends, New York State was beginning to use, or at least threaten to use, its regulatory authorities to address state pollution problems. In August 1965, the state health commission enacted a new law that could be used to bring action against the municipalities and industrial plants that were causing major pollution problems in New York. Solvay-Sennet was listed as one of the companies against which the state was threatening legal action.[21] In the lead up to the 1966 midterm elections, Congressman James Hanley began to take a serious interest in Onondaga cleanup efforts. Hanley, a liberal Democrat and strong supporter of Lyndon Johnson's Great Society, praised the Onondaga Lake Scientific Council on the floor of the House of Representatives for its "vigorous and sustained citizen response" to area water pollution issues.[22] The Council had recently released a study on lake algae content. It noted that algae had increased 10,000 times since 1962. The algae had been kept somewhat in check by industrial solvents flowing into the lake, which had increased its saltiness, but the algae problem was exacerbated when the decision was made not to build the Metropolitan sewage treatment plant as a tertiary treatment facility, and to halt efforts to divert the effluent into the Seneca River.[23] Oddly, the council also urged that a park be built on the Solvay wastebeds on the west side of the lake. Scientific investigation was in its early stages at this point, and it would be years before the full extent of wastebed contamination would be known. The park proposal had surfaced previously in 1953 but was never completed. As part of the proposal, several pedestrian bridges were constructed to facilitate access to the park, bridges that had subsequently been dubbed "bridges to nowhere."[24]

In the meantime, slimy wastes in Nine Mile Creek, the result of the relentless sewage dumping, clogged up the Solvay intake pipes, making it impossible to draw the six million gallons that it needed from the creek for industrial cooling purposes.[25] County Executive John Mulroy and Syracuse Mayor William Walsh announced that the county would take over the maintenance of another pipe, an eight-mile sewage line that ran along the lake carrying waste from the city's main lines to the treatment plant. Improperly maintained and cleaned, the line routinely overflowed, spilling raw sewage into tributaries and the lake itself. The move was hailed as "a step to end pollution in the lake."[26]

Congressman Hanley continued to play a role as the elected official most willing to support cleanup efforts. When the federal government earmarked $3.6 billion for federal water pollution control efforts, Hanley lobbied for Onondaga County to be offered a substantial proportion of the money for its pollution control efforts. Hanley invited Eugene Eaton, director of the federal Office of Water Resources Research, for a tour on Onondaga Lake. Afterwards, Eaton commented that he "needed a lot of soap and water to feel clean again." Hanley hoped to make Onondaga Lake a model of how federal cleanup efforts could be brought to bear on the nation's water pollution problems.[27] A new $500,000 study of the lake, partly spurred by the work of the Scientific Council, was announced at the end of the year.[28] This would constitute the best financed and intensive scientific analysis of the lake to date.

Onondaga's cleanup had now become an issue in the 1966 congressional campaigns. Hanley stated in one appearance that, if elected, he would seek federal funds to make it a "national example" of what could be done to redress national water pollution problem. His opponent, Stuart F. Hancock, a Republican, had his own "three point" plan to deal with area pollution issues. And Dr. Norman Balabanian, a candidate on the Citizens Peace Party and Liberal Party tickets, said that Onondaga Lake "must be renewed." Hanley worked hard to make the lake's cleanup a priority. In August, leading up to the November election, he convinced Alabama Congressman Robert E. Jones, Chair of the Subcommittee on Natural Resources and Power, to visit Syracuse to conduct a hearing on water pollution in New York State. Then Hanley announced that New York would be the beneficiary of a good proportion of funds from forthcoming legislation authorizing $3.6 billion for abatement of water pollution.[29] He would go on to win the race and serve for another fourteen years. Hanley also helped initiate another study, this one to examine the local interceptor sewer system, which carried rain runoff into the Syracuse Treatment Plant, and which created overflow during heavy storms. It had been a persistent problem that needed to be addressed if the lake's recovery was to ever be possible.[30]

All the publicity and intensifying political pressure appeared to have an impact on Allied Chemical, which in response, offered social scientists at Syracuse University a grant of $10,000 to study the "socio-economic aspects of pollution."[31] In November, in the midst of congressional elections and under pressure from the New York State Department of Health, both Crucible Steel and Solvay Process submitted a timetable indicating that they would cease pollution of Onondaga Lake by August 1, 1971. Crucible stated that it would stop dumping acids, sewage, and oily water into the lake by late 1969. Solvay Process promised treatment of its 4,000 tons of alkali waste, either in the Metropolitan treatment plant or its own facility (to be built). The agreement was described as a "turning point in years of citizen effort to clean up this lake in the heart of Metropolitan Syracuse," a lake that had by now been described by the Federal Water Pollution Control Administration as "quite possibly the most polluted body of water" in the Lake Ontario basin.[32] This would be among the first announced turning points. Many more would follow. It remained unclear, however, how sewage treatment would eliminate or reduce the massive amounts of calcium chloride and chemical contaminants, such as mercury and organic compounds, imbedded in them, that were an unavoidable outcome of the Solvay process.

The county was now under orders from an increasingly aggressive state government to upgrade the Metropolitan treatment facility.[33] At the same time, it was moving forward with plans for the 175-acre park, that was to be partly created by dredging Solvay waste sludge from the bottom of the lake.[34] This was a highly questionable idea, given that the sludge was contaminated with mercury and other toxic substances, but it was the first time that the idea of dredging the lake bottom was being proposed. Dredging would eventually become the centerpiece of lake cleanup efforts.

County Executive Mulroy announced yet another half million-dollar study. This time the aim was to survey the entire Ley Creek watershed, within which most of Onondaga's 135 industrial plants were located. One proposal was to convert the Ley Creek sewage plant into an industrial waste treatment plant. Like Hanley, Mulroy had become a leading elected official pushing for cleanup of the lake.

At the state level, Dr. Hollis S. Ingraham, the New York State Health Commissioner, had made pollution abatement a key part of his department's mission. Speaking at a workshop in Syracuse, he stated that the state's Pure Water Program had resulted in hearings at which "in almost all cases the industries involved have signed an order which legally requires them to end pollution by a certain date."[35] While laudable, the statement also evidenced a certain naivete in assuming that the polluting industries would abide by signed agreements, and that once they ceased polluting, the toxic legacies of pollution would resolve themselves. Ingraham, and no doubt other public officials as well, did not fully appreciate the long-term impact of contamination that had accumulated over decades. Moreover, while Ingraham had moral authority and a bully pulpit, the state health department had limited regulatory authority.

The Onondaga County Board of Supervisors announced an upgrade of the Ley Street sewage facility, which involved a partnership with Solvay and was intended to allow its treatment of some elements of Solvay discharges. It was not clear how this would work, given that Ley Creek is on the southeast side of the lake, whereas Solvay was on the southwest. A pipe would presumably have to be extended, bypassing the main Syracuse facility. Onondaga Lake has several tributaries, with Onondaga Creek and Harbor Creek being the two main ones, discharging 70% of the lake's inflow. Ley Creek, which accounts for 7% of the lake's inflow, acquired the nickname "Old Stinky" in the 1930s, because of the history of sewage and industrial effluent that was dumped into it. Along with the sewer upgrade, the county seemed to be moving toward dealing with the overflow problem from interceptor pipelines, that could be improved, it was believed, with a good cleaning. Part of the project also involved establishing a lagoon by the lake, which would act as a catch-basin, when during one of approximately forty-eight times a year, overflow would become a problem, sending raw sewage into nearby streams.[36] This was obviously not an ideal solution, since it sequestered the runoff within the lake itself. At the same time, the county authorized a bond to start development of the proposed park on Onondaga's shoreline.[37]

In June 1967, Dr. Daniel F. Jackson, a limnologist (i.e., a scientist who studies freshwater systems) at Syracuse University, delivered a paper at the Sagamore Lake Conference Center, in which he described Onondaga Lake as a "huge stinking mess of filamentous or blue-green algae."[38] Allied Chemical may have been starting to get the message, as it finally moved toward instituting some of its pollution abatement promises. The company committed itself to working with Onondaga County to build a multipurpose sewage facility that would deal with the pollution flowing into Nine Mile Creek, a tributary on the western side of the lake. While Allied was trying to bring Crucible Steel on board, they would go ahead without them, if necessary.

President of Allied Chemical John T. Connor argued that the Solvay Processing Plant had brought great benefits to the area in terms of investment and employment, but he recognized that there was some "resentment," given the waste that they had generated. Still, he noted, Allied had worked with the county to dredge Nine Mile Creek so that its detritus could be turned into a lakefront park. Moreover, he suggested that the public did not entirely understand the relationship of the Solvay process to the lake's condition. The company pulled water out of the lake at a rate of 100 million gallons daily, which it chlorinated to be used for its industrial purposes. After its use, the now sterilized water was returned "a good deal purer" than when it was taken out. He also stated that the discharges back into the lake were "not toxic or harmful" and contained "no bacteria or viruses." Rather, they were beneficial, he claimed, since they deterred the growth of algae.[39] He was drawing upon the Scientific Council's finding that algae blooms had been mitigated by chemical contamination. In other words, Allied was continuing to promote the idea that its effluents were not only not harmful, but actually good for the lake.

While Connor seemed to be trying to strike a conciliatory note, Warren S. Ferguson, technical manager at the Solvay plant, was complaining about complex and conflicting federal and state laws regarding air pollution. He argued that the rules designed to apply to new plant construction were not feasible for existing plants,[40] and went so far as to suggest that industry was simply being "scapegoated" for air pollution problems.[41]

Officials were now estimating that the cost of a "massive" new treatment facility that would prevent both industrial and municipal wastes from entering the lake would be $23 million, most of which would be paid for with public funds, with Solvay and Crucible Steel contributing about $4 million total. Thus, Solvay was off the hook for most of the cost of the plant, and plant managers could argue that they would meet their promise of ending the contamination of Onondaga by August 1, 1971. The wastes would be treated at the new facility, for the most part at taxpayer expense. What would happen to the massive amounts of calcium chloride generated by Solvay was not entirely clear. It would presumably be treated in some way and then dumped into the lake.

Air pollution continued to be an issue. In the Spring, Allied announced the planned installation of some electrostatic precipitators that would significantly reduce the amount of fly ash being released into the atmosphere. Fly ash was the result of coal burned in the soda ash production process and was generating noticeable localized haze. The precipitators would not, however, remove sulfur dioxide from the coal burning, the effects of which Warren Ferguson downplayed.[42]

Ferguson also suggested that Solvay might tap Seneca River water as a substitute for the water it was now pulling from Onondaga Lake. One reason was a rate increase imposed by the county, which would add $330,000 in charges on the company. But ironically, the company was also concerned about the water's quality and whether it was even suitable for industrial purposes, given its high levels of contamination. Seneca water was hardly pristine, but it could be suitable if treated with chlorine. The company held a right of way for a pipeline to transport the river water back to the plant, but it would be an expensive proposition to run a pipe ten miles from the Seneca River back to Solvay. Given the choices, Ferguson suggested that Solvay might have to leave the area. "If it gets to the point where the lake water is that bad," he stated, "I think it also would be a question of how long our firm is in New York."[43] Ferguson's threat was the ultimate irony: the Solvay plant, which was a main source of Onondaga Lake's pollution, and which encouraged the dumping on sewage on its wastebeds, might leave Onondaga County because Onondaga Lake's waters had become too polluted for it to use.

Meanwhile, plans for an adequate municipal sewage system dragged on. Syracuse had no choice, it seemed, but to continue mixing its sludge with Solvay process material until something was done. The Mayor's office released a statement: "Years ago, with the approval and under the supervision of the State Department of Health, the city erected a plant and perfected a process whereby sewage from the city was mixed with waste from the Solvay Process Company." In retrospect, "perfected" seems like an odd choice of words, but that is how both city and company officials perceived it, given that the main purpose of the mixing was to keep the smell within tolerable limits. However, the mixing process did nothing to reduce the toxic substances that were in both the Solvay waste and the sewage sludge.

Now that the company had moved its active wastebeds because the previous ones had filled to capacity, the city's only alternative was to dump its raw sewage directly into the lake. Given the seeming incapacity of the city and surrounding towns to get an agreement on building a new sewage plant, the mayor suggested that it was time to place a new set of pipes that would reach over to the relocated Solvay beds. "The pipeline," the mayor argued, was temporary, initiated at the request of the Department of Health. Ultimately, "The mixture of Solvay sludge and the city sewage would go into exactly the same beds as it has for the last few years."[44] The mayor voiced commitment to a new sewage plant, but the city and county seemed to be stuck on the same old treadmill.

A lakeside park, an idea first floated in 1954, was again proposed. As part of the plan, a dike would be constructed adding more than fifty acres of filled land to the 175 acres of marshland (what we would today call "wetlands") that would become the earthen foundation for the park. The Nine Mile creek delta would be dredged of its calcium and sludge deposits, and the 650,000 cubic yards of muck resulting would be used as fill, on top of which would be added about 500,000 cubic yards of earth. This operation, while it would give area residents an opportunity to revisit the lake front, would do little the improve the condition of the lake itself, in fact, it created an additional problem in that contaminated groundwater could now seep back into the lake, picking up more contamination after being filtered through the dredged material.[45] A version of this park was eventually completed, but it would take another 40 years for that to happen.

The 1970s: Opposition Intensifies

In April 1970, as the country celebrated its first Earth Day, governor of New York Nelson Rockefeller signed legislation to create the New York State Department of Environmental Conservation. The DEC would become an important actor in state politics and would play a crucial role in eventual restoration attempts of Onondaga Lake. And while the Department did not have an immediate impact on point source pollution, either from industry or municipal sewage, its existence did indicate that New York State would begin the process of establishing a regulatory regime.

Syracuse Mayor Ben Walsh had promised to build a new sewage disposal plant while in office, but as of spring of 1970 when a new mayor, Democrat Lee Alexander, entered office it had not been constructed. A proposal was on the table for a $48.8 million facility, but only four members of the public showed up at a hearing on the proposal, two of whom opposed the project. Syracuse University engineering professor Dr. Daniel Jackson, suggested that the plant wouldn't lead to any improvement in the Onondaga Lake water quality, and that similar results could be gained from simple modifications of the two existing plants.

William Maloney, a long-time activist committed to lake cleanup, objected to allowing waste from the Solvay facility to be processed by a municipal plant. "We aren't obliged to Allied Chemical," he stated, "It's their problem." But John J. Hennigan, the assistant director of public works, countered that mixing Solvay process wastes with sewage would benefit the system by "destroying phosphate," a task that would have to be done one way or another and could be costly. The city was still committed to the mixing of Solvay waste and municipal sewage, contending that it would result in the two neutralizing each other. Hennigan asserted that the new facility would significantly improve the lake's water quality, making it available for recreation, including swimming. None of this was true, and, moreover, the new plant would not resolve the seemingly intractable problem of influx from storm sewers, a legacy of a decision made in the early twentieth century to combine storm and sewage inflows. As a result, during heavy rain events, overflow would receive only primary treatment, which amounted to mixing it with chlorine. Finally, most of the costs of the new plant would be borne by county residents and the state, since promised federal contributions had now declined significantly.[46]

As Allied Chemical entered the 1970s, significant changes were on the horizon: company president and Syracuse native John T. Connor raised the possibility that the "synthetic" process of producing soda was being superseded by what he called the "natural" process of extracting it from soil, which was applied at Allied facilities in Wyoming. "Certainly," he noted, "there will be no expansion of the soda ash production by the synthetic method." In other words, the Solvay process, which had been the chief method for soda production for a century, was beginning to lose its primacy, and this would no doubt at some point, impact production in Solvay. Moreover, one of the primary uses of soda ash was for glassmaking, and glass production was declining due to the proliferation of plastic bottles for soft drink storage and delivery. In an attempt to diversify and seemingly oblivious to the irony of such a move, the company was now considering entering the market in pollution control devices.[47]

Allied also continued its efforts at public relations. It put together a "fast-moving sound slide story," designed to illustrate how its settling beds operated, a "misunderstood" process, according to them, which many in the public incorrectly believed resulted in their expelling waste

materials into the lake. The show was replete with a miniaturized dem-
onstration of how the Solvay process worked using laboratory flasks.[48]

While the company was hinting that the Solvay plant could be replaced by mining operations in the west, it was also seeking to expand
the quarry where it had been accessing the limestone that was a primary
constituent in producing soda ash. But it would now have to deal with
NIMBY advocates, inspired by the emergence of a widespread environmental awakening. Residents in potentially affected areas were offering
resistance. Suburbs now ringed the City of Syracuse, resulting in an expanded population of residents fiercely protective of their properties and
communities. Opposition to quarry expansion was growing in the Town
of Manlius, where residents were concerned about increases in truck
traffic, impacts of blasting operations, contamination of well water, and
reduction of property values. The quarry had been in operation since
1911 but had been contained within the Jamesville town limits. With the
proposed expansion, the Manlius Planning Board would have to give
its approval, so local residents actually had some clout. Moreover, residents' economic fortunes were not directly tied to the company, insulat-
ing them from the economic threats that had worked so well for Solvay
in the past.[49]

Legal challenges to the plant's water discharges were now being initiated by the federal government. U.S. Attorney James M. Sullivan announced a lawsuit against Allied to immediately stop the discharge
of mercury by the facility into Onondaga Lake. The lawsuit, brought
at the behest of Richard Nixon's Interior Secretary, Walter Hickel, was
grounded in the rarely enforced 1911 Refuse Act, which was designed
to prevent any materials other than sewage from being discharged into
navigable waters of the United States. Solvay's Director of Operations,
James Campbell, responded that while the company had not been able to
entirely eliminate mercury discharges had done much to reduce them by
diverting them to its wastebeds, where they were "locked into" calcium
deposits. He also claimed that mercury was only recently discovered to
be a contamination problem, because previously it had been assumed
that the heavy liquid metal sank to the bottom where it was rendered
essentially harmless.[50]

The lawsuit generated a crisis at the plant when Republican Governor
Nelson Rockefeller joined with the federal government and threatened

a complete shutdown of its operations if it did not comply with an order to stop mercury discharges. In response, Campbell promised that the company would do its best to comply with the order, "not only because it is a social responsibility we readily accept but because we want to protect the future of our plant in Solvay and the jobs of our 2,300 employees."[51] In response, Allied submitted a plan to the state that prevented the shutdown from going forward and staved off the crisis, temporarily at least.[52]

In 1971 New York approved a state grant for a new Metropolitan Sewage Treatment facility. Supplemented with a federal grant, it would cover more than half of the cost of the facility, whose costs had now risen to nearly $46 million.[53] While the state's contribution was welcome, it would not be adequate to provide for the needed tertiary treatment plant.

The town of Manlius's attached twenty stipulations to its decision to accept a proposed Allied quarry expansion. These included controlling air pollution, noise pollution, appropriate fencing, and limits on hours of operation. The company claimed that it had a right under its original permit to expand its operation to Manlius, because it was approved before Manlius had instituted its zoning ordinance, so the expansion should have been grandfathered in. But the town Planning Board seemed prepared for a fight.[54] Local residents were legitimately concerned that the additional 2,500 acres of land used for the quarry would not only generate noise, traffic, and pollution, but would also prevent the town from developing as a normal suburban community.[55] After all, what developer would want to build housing near a stone quarry, and what homebuyer would look seriously as such properties if others were available?

By the spring of 1972, in spite of continuing promises to do so, air pollution generated by Allied, Crucible Steel, and Alpha Portland Cement Plant had not been mitigated. Allied was still burning coal for electrical generation to run the Solvay Process facilities. While the company claimed that it had reduced its pollution levels by 95%, some of the process stacks were still out of compliance with state law. Crucible Steel claimed that the placement of a bag house (a kind of filter, developed in the 1970s for industrial plants) would significantly reduce its emissions of ferric oxide, which would improve the county's air quality. Alpha Portland Cement promised that it would have a precipitator in place by fall. Suggestions that the facilities be closed if they continued to cause air pollution problems were dismissed out of hand, given the 3,500 jobs at stake, a $40 million payroll, and millions in taxes paid by each the facilities to local governments.[56]

In August 1973, the Manlius Town Board unanimously rejected Allied plans to expand its quarry operations, representing a continuing and profound shift in local attitudes toward industrial expansion. The basis for the Board's decision was nominally that Allied had not submitted a reclamation plan for the 46 acres in question. The Board also determined that "sufficient unquarried lands" were still available for Allied's operations to continue for several years, thus making expansion unnecessary.[57]

Onondaga County was using carrots along with sticks, to encourage Allied to reduce its discharges. In 1974, the Onondaga Industrial Development Agency said it would offer $30 million in bonds for pollution control measures for Allied early in the year. The bonds would cover projects in all five of the company's New York State operations, and would cover "heavy metal effluent discharges, control of spills and chemical leakage" as well as fly ash emissions. The OIDA asserted that this would provide funds sufficient to cover each of these areas, with room to spare.[58]

In spite of its various problems, in the spring of 1974 Allied was predicting a robust financial future. Profits were up even from the previous year's "record performance." Soda ash was in demand, and the company could not produce it fast enough. It was increasing investment in its Wyoming operations, which involved mining rather than factory production processes. In addition, the company was developing polyester fabrics, which were made into dresses worn at the announcing event by the hostesses, each sporting a design from Allied's recently retooled logo.[59]

At the same time, the company was coming under increased pressure from state regulators over its air emissions. In August, the state Environmental Control Agency held an enforcement hearing, because several of Allie's boilers were out of compliance with state fly ash emission, and the company had agreed to come into compliance by July. The state wanted the company to install precipitators to control the fly ash, but Allied officials were reluctant because the state's proposal would require complete replacement of all the precipitators at considerable expense. The company wanted only to "upgrade" them. It further contended that the problem could be resolved by burning better quality coal. High grade coal had been difficult to secure due to the energy crisis, but the company was making moves to resolve that problem by improving its distribution network. The state was not threatening coercive action such as fines, at least not yet, but it was demanding good-faith compliance.[60] In fact, the Onondaga County Industrial Development Agency continued to offer financial incentives in the form of a $10,000 bond to Allied to improve is pollution control measures, yet to be taken advantage of by Allied.[61]

In spite of state and federal funds being available, the sewage treatment issue remained unresolved. The Sierra Club objected to the plan for tertiary treatment, which involved using Solvay's calcium chloride to react with phosphorus in the sewage as chemical removal agent. The result would be precipitation of calcium chloride at the south end of the lake, which had previously been free from it, and the material would have to be dredged. Moreover, the use of Allied's settling basin as a location for the deposit of sewage sludge continued the unholy alliance between public authorities and the company and could be used to justify expansion of the wastebeds into the town of Camillus, now with the support, apparently, of city, county, and state governments.[62]

Allied had also been asking for a zoning change in the town of Camillus in order to expand its wastebeds and to increase their depth to seventy feet. As in the town of Manlius, however, resistance was building among residents. And now, Allied's threats of leaving the community were working against it. As one prescient letter to the editor of the Post-Standard noted, commitments regarding restoration of land used for such purposes needed to be secured up front, because of the possibility of the "eventual termination of local Allied operations for whatever reasons it might be."[63]

If 1974 started out looking good for the company's balance sheets, by the end of the year things had taken a turn for the worse. In the middle of the economic crisis popularly referred to as "stagflation," demand for soda ash was down, and given the then current economy, the company was facing negative growth. The plastics division was the only bright spot.[64]

In 1976, the federal EPA questioned Allied's compliance with federal water pollution regulations because of the heat pollution generated by its discharges from the Jamesville quarry. Allied challenged the EPA findings by demonstrating that trout were thriving in the stream, which led to the EPA withdrawing its objections.[65] Despite that small victory, Allied continued to be dogged by issues related to its discharges of ketone into the James River in Virginia. A federal grand jury put forward an indictment in May 1976. The company was already facing a $100 million lawsuit from company employees who had been exposed to the chemical and had experienced various health problems, including tremors, eyesight damage, and increased cancer risks. Ketone had been used as a pesticide, and its release into the river had significantly damaged the local fishing industry.[66] In October 1976, these practices moved a judge in Virginia to levy a $13.2 million fine against Allied Chemical. In the judge's words, pollution "is a crime against every citizen."

By summer of 1977, a much more optimistic view of Onondaga's potential was starting to emerge, prematurely, it needs to be said. In an article in the *Herald American* newspaper titled "Onondaga Lake Lives Again," reporter Michael Kelly asked, "Will Onondaga Lake be the county's most popular recreation spot in the 1980s?" "Don't laugh," he wrote, and added that the lake now "teems with game fish." Moreover, the completion of the Metropolitan Sewage Treatment plant, involving a public-private partnership, would remove the most egregious pollution problems, providing "tertiary" sewage treatment. In addition, the county was moving forward with a plan for bicycle and hiking paths, picnic facilities, and a marina. Electrically powered trams would move visitors through the park, so that automobile traffic could be kept to a minimum. The Solvay wastebeds would become part of a park system in what Parks Commissioner James R. Johst described it all as "the most exciting project the county has undertaken in many years."[67]

Still, opposition to waste disposal of any kind of was intensifying in the Syracuse suburbs. In 1977, The State Department of Environmental Conservation and County Solid Waste Disposal Authority had determined that the best site for a regional landfill was in the town of Camillus, next to the current Camillus dump, on land owned by Allied, sitting on an already depleted part of their limestone quarry. Allied agreed to turn over the land to the county, but fierce opposition emerged almost immediately, with area residents and political leaders up in arms after the announcement was made. Town of Camillus Councilman Leland

Blanding asked, "Is the SWDA being managed with any brains at all?" Also in June of that year, some county residents turned up at a meeting of the Solid Waste Disposal Authority to express objections to thirteen acres of Allied's Jamesville Quarry being used as a landfill. Allied had actually already agreed to drop the idea, given public opposition. But just in case, a number of locals appeared with an engineering firm claiming that the facility, if pushed forward, had the potential to contaminate groundwater. Even a spokesman for Allied, which owned the Camillus property, suggested that groundwater in the nearby town of LaFayette could be contaminated if the landfill was located there.[68]

Another site being considered was south of Syracuse in the town of LaFayette, a quiet, upscale suburb dotted with private swimming pools and horse stables. Residents there were also quick to mobilize and challenged SWDA's plan even to conduct bore tests at the proposed site. One resident noted, "We've lived in a paradise, and now we're being put into a garbage pit."[69] While this controversy unfolded, contrary to the explicit orders of the state DEC, the county was continuing to dump its refuse at a town of Clay site despite the existing overflow.[70]

Allied's proposal to expand its quarrying into Manlius, which had been in the works for years, continued to inch forward against tremendous opposition and a decision on the part of the town to reject the needed permit. At a hearing on April 1978, residents expressed their displeasure, and they came armed with legal arguments. Zoning was a key issue. The land had been zoned agricultural when Allied had established its original quarry, and for the quarry to expand, zoning needed to be changed. The company originally requested agricultural zoning because the designation allowed the corporation a significantly lower tax rate over many decades compared to an industrial zoning. As a result of the zoning, houses were built on the assumption that the area would remain free from industrial development. Now Allied was trying to change rules from which it had benefited but which had also created certain expectations. The problem was of Allied's own making. The company tried to alleviate concerns by offering guarantees that it would reduce noise levels from blasting and moderate traffic, create buffers, and reclaim the land once the quarry was exhausted. Such promises did not satisfy area residents.[71] After all, this was a company that had a long, established history of generating intractable contamination problems.

The bottom of the bridge on Darwin Avenue that crossed Onondaga Creek had become a popular fishing spot for young boys, since the creek's water quality had improved over the years. Not only were there plenty of carp, but trout as well. The boys would throw the carp back, but keep the trout, catching fish in the range of twelve to sixteen inches. But in the summer of 1979, a pipe leaked and poisonous brine from the Allied facility spilled into the water, causing a massive fish kill in the creek.[72] Crayfish, which had also been plentiful, were now gone as well. A local state assemblyman proposed that an alarm be installed to warn when future spills occurred, and even offered to help find a state grant to pay for it.[73] But even before the fish kill in June, the level of public opposition to reopening part of the Jamesville Quarry faced so much opposition that in April of 1979, Allied decided to pull the plug on it.[74]

Rumors continued to swirl around whether the Solvay Plant would cease operation entirely. An article appeared in *Business Week* suggesting that the plant might close within a year. Allied denied the report. While mining of trona to produce soda ash was less expensive than the Solvay process, Allied contended that transporting it significantly increased its costs, making it impractical as a complete substitute.[75]

The 1980s: The End of an Era

As Allied entered the 1980s, the New York Department of Environmental Conservation announced significant improvements in the air quality surrounding the Solvay Plant. The company had installed electrostatic precipitators on its 100 smokestacks as well as a dust collection system to remove limestone dust that it generated. It had reduced micrograms of precipitates from 103 per cubic meter in 1979 to fifty-six micrograms in 1978. These changes now met state health standards, but not its "aesthetic" standards. Residents had mixed views of how much improvement had been achieved. A janitor at the facility noted that there was now much less dust inside the plant, and the proprietor of a variety store nearby said that he no longer had soot in his swimming pool. Others were less impressed. One resident stated, "The air around here hasn't improved at all." Another complained, "You can tell them I said it still stinks."[76]

Allied's desire to expand its quarry operations continued to provoke controversy. The company was threatening to sue the town of Manlius if it denied the company a permit for the facility. Public opposition, however, had not abated. When one opponent suggested that the company was soon going to leave the area, Allied attorney Earl Evans said that it had no plans to do so and promised that it would stay in the community "as long as the natural resources were available."[77] It appeared to be a veiled threat.

1981 marked 100 years that the Solvay Process plant had been in operation. Much had transpired, of course. While managers might be pleased at the longevity of the facility and its contributions to the economic and social life of the local community, the long legacy of contamination—the potential for which was never given any attention in Ernest Solvay's original vision—had now become a signature part of its reputation. The company had gone from being a highly valued economic asset that had brought jobs and industrial progress to the area, to being a source of toxic contamination, using its clout to blackmail area residents and protect its bottom line.

Residents of the town of Manlius had lost their fight to prevent Allied's limestone quarry from being expanded into their community. In spite of company promises to keep its disruptions to a minimum, residents complained bitterly about the noise and danger posed by blasting activities. The company defended its environmental record, however, with a spokesman contending, "we have really done a good job in cleaning up. We've been better than most." But the question of how long the plant would stay open continued to linger. It was by now the only operating soda ash plant left in the U.S.[78]

In March of 1983, the Onondaga County Legislature issued another in what was a continuing stream of reports on the state of the lake. The lake had now become, by some estimates, not only the most polluted lake in America, but also the most studied. The reports noted marginal improvements in some aspects of water quality, but raw sewage was still being dumped in, and mercury continued to contaminate the water, even though Allied's practice of disposing of it in the lake had ended. Ironically, as calcium chloride waste in the lake declined, it resulted in the release of methylmercury into the lake itself. The report's conclusions were obvious. Much more needed to be done before Onondaga Lake could again be considered one of the jewels of upstate New York.[79]

In July 1984, the state DEC charged Onondaga County's Metropolitan Sewage Treatment Plant with releasing an unacceptably high level of solids into the lake. The county denied the violation, but it seemed clear that the solids involved were bits of calcium compounds that resulted when the county mixed the Solvay plant's calcium chloride with the county's sewage to remove phosphorus from it.[80]

In December 1984, Allied sold off a piece of its production facilities. Church and Dwight, Inc., which owned the firm Arm & Hammer, known for its bicarbonate products, bought that part of the facility and stated its determination to continue operating it at its current employment levels.[81]

In spite of Allied's efforts at spin, it was now clear that the legacy of its pollution would persist for years to come. Environmental activist Walter Hang, a specialist in molecular biology with a focus on the toxicology of pollution, was working on a study that showed high cancer rates in Onondaga County linked to the pollution in Onondaga Lake. The County was in the top 10% of cancer deaths in the state.[82]

In general, environmental issues in the early 1980s were essentially in a holding pattern. No significant progress was made toward moving the Metropolitan Treatment Facility toward tertiary treatment. Overflow problems persisted. Allied had stopped dumping its calcium chloride deposits into the lake, but these were now being pumped into settling ponds near Nine Mile Creek. Since these were no longer being infused with the lake water to react to the phosphorus from the county's sewage discharges, and since no pipe had been laid to transport the sewage to the wastebeds, the lake became ever smellier.

In April 1985, after rumors had been circulating for several months, Allied announced that it would shutter the Solvay plant. 1,100 of the plant's 1,400 workers, with an annual payroll of $46 million, would lose their jobs. Soon the small industrial city's streets, once teeming with workers on their way home after a day's or night's shift, would become deserted. Restaurants and taverns that had served Solvay workers would become empty, and many would eventually close. The city would lose $3.4 million in tax revenues and $150,000 in annual charitable donations. It was the end of what one employee described as "a way of life."[83] Solvay, New York, faced a crisis of deindustrialization, not unfamiliar to many other cities and towns in upstate New York and across the Great Lakes region.

Why did the plant close? It might be tempting to propose that the continuing pressures from environmentalists and public authorities to address its pollution problem played a role in the decision, but they do not seem to have been central. The product most in demand for soda ash, glass, had been in decline for years, displaced by plastic and aluminum soft drink containers and a recycling process for glass making that did not rely on soda ash. The Solvay plant had been operating at a loss for three years. Moreover, it was an old facility and not suitable for any other purpose. Virtually all of the buildings would be demolished.[84]

While the economic impact of closing the Solvay plant was traumatic, it also generated a renaissance of renewed interest in Onondaga Lake. A *Syracuse Post-Standard* feature article entitled "Onondaga Lake: A Paradise Lost?" recounted environmental problems with the lake and detailed some of its history. Among their findings was that Solvay Process was far and away the lake's worst polluter. "Since the 1880s, Allied has literally been filling the lake with waste—tons of sodium chloride and calcium chloride." According to the article's estimates, somewhere between 20 to 40% of the lake had been filled with Allied wastes, which had now become a messy goop on the lake bottom. Among the chief contaminants were mercury (twenty pounds a day for eighteen years). Levels of mercury had dropped briefly but were now back on the rise. Allied was not the only offender, however. Other industrial concerns had also contributed to the lake's contamination, including Crucible Steel, which contributed chromium, and Bristol Meyers, which added wastes from its penicillin operation. And of course, the City of Syracuse had been dumping its sewage into the lake for 100 years, with raw sewage still flowing into the lake during periods of heavy rain.

The article described collusion between government officials and the polluters themselves. In the 1920s the city began to use the lake as a sewer, on the grounds that "sewage would decompose entirely as it went from one end of the lake to the other." Also, "state officials failed to enforce state law against lake pollution. Instead, the city and the state made quiet deals allowing Allied to continue despoiling the shoreline and to continue discharging waste into the lake." Politicians such as Governor Thomas Dewey, who made promises to clean up the lake, never fulfilled them, and he blocked any legal action against Allied Chemical. Local newspapers also tended to ignore the problem. An article in the *Post-Standard*, written in 1903, attempting to describe some pollution problems, was "remote from the truth."[85]

At the same time, however, a "handful of heroes" attempted to call attention to the lake's decline and to turn it around. These included SUNY College of Forestry and Environmental Sciences Professor Norman Richards, who spent hours of his time in the sixties and seventies lugging seeds and fertilizer to the Solvay wastebeds in an attempt, eventually successful, to cover them with grass, shrubs, and wildflowers. Walter Welch, Crandall Melvin, and William Maloney formed a trio in the 1940s and 50s that was the first group to push for cleanup, "well before the environment was a hot issue." And finally, civil engineer Daniel Jackson and chemist Samuel Sage were constant critics of county officials. Sage used his boat, the Saltine Warrior, to build support for lake cleanup.[86]

A turning point in environmental consciousness occurred on Thanksgiving Day 1943, when the dam holding back the Solvay wastebeds broke, and an "eight-foot wall of goo, thicker than week-old mashed potatoes, flooded into the Lakeland neighborhood." A square mile of the area was flooded, and residents started to ask, for the first time, it seemed, "what Allied was doing to the lake and its shore." The state eventually reached a deal with Allied and agreed not to press for damages resulting from the Lakeland spill, in exchange for Solvay's giving the land where the wastebeds were located to the county.[87] Over time, however, as scientific investigations probed ever deeper into the mysteries of Onondaga Lake pollution, findings would reveal the extent and decline and the complexity of the threats posed to it. In spite of it all though, many believed that with Allied's departure, the lake's future was "bright."[88]

Jackson and Sage wanted to build the Metropolitan Sewage Treatment plant at the north end of the lake, on the grounds that the faster moving water would help to decompose any remaining organic wastes. Commissioner John Hennigan, however, supported the idea of mixing Solvay's calcium wastes with the sewage, a proposal that environmentalists disliked, because it legitimized Allied's calcium contamination. Hennigan won the battle, and denigrated Jackson, calling him not a civil engineer, but a "bug worshipper." Hennigan had also predicted in 1979, that the lake would be cleaned up within a couple of years of the operation of the new sewage treatment plant, an estimate that he eventually admitted was too optimistic.[89]

The contamination problems of Onondaga Lake were so widespread and interrelated, it was difficult to disentangle them. One example of the complexity of the lake's ecology involved analysis of its high salt content. Allied officials suggested that the lake was naturally salty. This was, after all, an area geologically rich with salt deposits. Robert Hennigan, a faculty member of the Environmental Sciences and Forestry School and "the dean of water quality experts" (not to be confused with John Hennigan), determined that the excess salt in the lake was surface salt, and thus clearly the result of discharges from Solvay. Moreover, because calcium dumping into the lake had ended, phosphorous levels would rise temporarily, because they would have nothing to bind with. The result would be intensified algae blooms that would add to the eutrophication (decline of oxygen content) of the lake.[90]

In October 1985, the DEC announced preliminary plans requiring Allied to participate in the lake cleanup, which would involve a study of the groundwater under the wastebeds, a determination of how much mercury was on the lake bottom, an analysis of chemical contamination at the Willis Avenue tar pit, the preservation of the caves at Jamesville, which were home to an endangered species of bats, and a plan for dealing with asbestos in the buildings that had housed the Solvay Process works, which were going to be torn down.[91] This marked the start of the DEC's decision to hold Allied accountable for Onondaga Lake pollution.

Allied was under pressure not just in Syracuse, but other places where it had left a toxic legacy. Their insecticide plant at James River in Virginia had become notorious, but there was also toxic contamination at a chrome plant in Maryland. Nevertheless, an Allied spokesperson continued to insist that the company would "Do what's right."[92]

In November 1985, Allied, rather than committing to any kind of cleanup of the mess that it had created, announced that it might buy the land to continue to operate its wastebeds. The city of Syracuse even approved of this plan because this would allow it to continue dumping its sewage sludge there, and it would save 100 jobs involved with the site's maintenance. The company contended that the beds could be used for the disposal of the sludge for fifty years. The New York Department of

Environmental Conservation itself seemed amenable, even though the wastebeds were leaching large amounts (1.6 million pounds) of chlorides into Six Mile Creek. Perhaps, the company suggested, Allied calcium chlorides could be mixed with brewery sludge, then being dumped from an Anheuser-Busch facility, to make a fertile compost that could be used to grow trees and other plants.[93]

The Post Standard reported in early 1986 that, in spite of the closing of the Allied facility in Solvay, construction permits in the western suburban towns of Geddes, Camillus, Marcellus, and Stafford were either on the rise or holding steady. These permits were for both business and residential purposes. It seemed that perhaps the Syracuse area economy would be able to weather the loss of the Allied facility.[94] Eventually this steady suburban growth would come into conflict with lake cleanup efforts, but that was a decade away.

In early 1986, another study was proposed by Onondaga County to determine what the impact of ceasing the introduction of Allied Chemical chlorides into the waters would be. Ordered by the state to reduce ammonia discharges into the lake, the county was hoping to demonstrate that such reductions would not actually improve the lake's ecology. The argument was that the ammonia would react with the lake's calcium chlorides to diminish the impact. But phosphorus represented the wild card because there was a fear that without the chlorides, with which the phosphorus bonded, large algae blooms would again turn the lake into a festering stinking mess.[95]

The other problem was what the county would do with the sewage sludge that it had been dumping in the Solvay wastebeds. It was unclear what would happen to the material if the beds were closed. One option was to continue dumping at the site, which was still being used by Anheuser-Busch and Linden Chemical, Inc., to deposit their wastes. Also, Summit Industries was proposing to buy Allied's power plant and use the wastebeds to deposit its fly ash. Considerable resistance was offered against this idea, however, because it would essentially let Allied off the hook in terms of cleaning up the site, not to mention the fact that fly ash is a highly toxic substance.

The DEC was indicating that it would only offer a two-year permit for such dumping, and in the meantime, the county would have to obtain a permit from the town of Camillus, where the beds were located.

Other options for disposing of the sludge included using the old disposal beds that had been closed for a number of years, closer to the lake and Nine Mile Creek or drying the sludge and having it hauled away (at a cost of around $4 million a year).[96] Another plan that was being floated, especially by Republicans on the county legislature, was to turn Allied's power plant into an incinerator that would burn trash and sewage sludge, with the power generated being sold to Crucible Steel and LCP Chemical nearby.

County Executive John Mulroy was unenthusiastic about the incinerator project, but agreed to find funds for a preliminary study.[97] However, a competing proposal by Hydra-Co would turn the plant into a traditional coal-fired facility that would include cogeneration of steam, providing eighty megawatts of power, enough to power the entire city of Syracuse and provide steam for local industries.[98] The state was mandating that state electricity generators include cogeneration as a part of their mix. Hydra-Co was a subsidiary of Niagara Power and would help to satisfy its requirements for cogeneration under state regulations.

The Hydra-Co proposal, however, would run into opposition from residents and the town board in the town of Geddes, which had begun to appreciate the clean air that they had experienced with the closing of the old Allied power plant. John Ferris, an attorney representing the town, stated that implementing the proposal would "make the entire Iroquois Confederation weep," with its contributions to local pollution and the sulfur dioxide that would travel as far as the Adirondacks. Advocates argued that it would meet the strictest state standards for pollution control and would be a vast improvement over the Allied generating facility.[99] While backers of the plan, which they called Salt City Venture, had hoped to expedite the process by avoiding a full environmental impact statement, the state ultimately decided that it would need to complete one.[100] Moreover, the public service commission began to question whether the proposed facility met state requirements for proportion of cogeneration to conventional power generation.[101]

The plan to continue dumping at Camillus ran into considerable opposition not only over concerns about contaminants leaching into Onondaga Lake, but also the impacts of odors on area residents. The county was leaning toward dumping the sludge onto the old wastebeds near Nine Mile Creek and on the shores of the lake itself. The temporary solution would involve lining the old beds with plastic to prevent leakage and would continue only for a period of six months, until the new sludge drying facility was completed. This plan was being pushed forward by acting drainage and sanitation commissioner John Karanik.[102]

By the fall of 1986, Allied reached an initial agreement with the DEC to study the content of the 1,000 acres of wastebeds that populated the area, and to determine what substances were actually leaching into Onondaga Lake. Many assertions had been made and conclusions drawn, but no solid evidence actually existed in terms of the specific chemical composition of the beds.

The primary waste product was believed to be calcium chloride, an inert salt, but mercury and asbestos were also likely embedded in the material. Given the quantity of material involved, thousands of tons, it was unlikely that it would or could all be removed entirely. But once the extent of the problem was determined, then presumably barriers could be constructed as needed, to minimize further leakage.[103] At the very least, it was now abundantly clear that substances leaching from the wastebeds were posing a considerable impediment to improving the lake's quality.

Steve Effler, a scientist working for Upstate Freshwater Institute, had been doing daily tests for more than a year and found that the lake was severely oxygen-depleted, especially the bottom twelve to fifteen feet. Effler's findings ran contrary to the widespread belief that the lake would simply start to heal itself as Allied's effluent discharges into the lake ceased and as fresh water from tributaries flushed out the lake. Effler's analysis suggested that direct contamination from Allied's dumping was now being replaced by materials leaching from the wastebeds. Given their enormous size, if left untouched the wastebeds' impact on the lake would continue into the foreseeable future.[104]

In fact, the optimism about the lake's future that had been expressed when Solvay Process closed, was now being replaced with a more realistic and pessimistic appraisal, as evidenced by an article in the *Syracuse Herald-American* that asked, "Can It Be Saved?" Steve Effler's analysis referring to the "unique combination of degradation" the lake had been subjected to featured prominently in the piece. Salinity was now recognized as a major problem, and calcite deposits along the shoreline had created a "literal zone" that prevented the growth of plant and aquatic life. Sewage outflows continued, and contaminants were still leaking into the lake from the Allied wastebeds.

Federal money was available for upgrading sewage treatment facilities, but in order to qualify for it, districts had to show that they could achieve a tangible benefit, such as a return to fishing or swimming in a previously affected body of water. The multiple pollution problems in Onondaga Lake made this difficult to demonstrate. The closing of the Allied plant, it turned out, was hardly a panacea, because of the quantity and content of wastes that had been deposited into the lake for decades. Effler noted, "If you could take away 100 years of history and start with a new lake, you put in the materials that we continue to put in and you'd have a lake with serious problems in a matter of years."[105]

Yet, in spite of these setbacks, the lakeshore was being touted as a potential tourist area. In August 1987, a "Waterfront Extravaganza" was held, replete with fireworks and other activities "that are common on most lakes—yet sadly rare for Onondaga." The seventeenth-century Sainte Marie Gannentaha Fort was being renovated,[106] and a new mall was proposed for the south end of the lake. Proposals were being floated to suck the goop from the bottom of the lake, but a sewage system upgrade still seemed like a fantasy, given the now $170 million proposed cost. Hennigan estimated that it would be at least ten years before construction could take place.[107] At around the same time, the DEC released a report showing that mercury levels were increasing in the lake. In a document entitled "An Overview of Mercury Contamination in the Fish of Onondaga Lake," DEC researchers found that from 1970 to 1979, mercury levels had inched upward. However, the source of the contamination was not clear.[108]

In September, the DEC granted tentative permission to reopen the Allied power plant as a cogeneration facility, but opposition flared up from local environmental groups, and even from within the agency's division of fish and wildlife. Of particular concern was the lack of scrubbers. While the cogeneration facility operation would emit considerably fewer particulates than the old facility, it would still emit nearly 13,000 tons a year of sulfur dioxide.[109] A public hearing brought out angry local residents.[110]

The county still faced a "sludge crisis," as it was becoming clear that "dewatering" the sludge and having it hauled away was an expensive proposition. Officials were hoping to get permission from the state and the town to return to dumping at the wastebeds.[111] The latest proposal for dealing with the sewage sludge was to mix it with lime, a practice used in some facilities in Ohio, but the proposal was met with skepticism from Camillus residents and the DEC. Nevertheless, even as the pollution problems proved to be increasingly intractable, plans were being submitted or major development projects on the lake.[112] The vision of the lake as an economic asset would continue to be a major factor in proposals to return it to some semblance of its former condition.

As 1988 began, the New York Department of Environmental Conservation gave LDP Chemical permission to dump its effluents into Geddes Brook, one of the small tributaries feeding into Onondaga Lake. The company asked for temporary permission to do this because the line that had brought the waste to the Camillus wastebeds was frozen, and, if the company had no way to expel its wastes, it would have to shutter the plant, perhaps permanently. The economic rationale was clear, but the environmental one was troubling. The DEC's dubious justification was that with so much contamination leaching into the lake from Allied's wastebeds anyway, "a few thousand pounds more couldn't do too much more damage." The discharges included "calcium, sodium chloride, and sulfates as well as some mercury." The DEC contended that, given the 800,000 pounds of calcium compounds that were already seeping into the lake every day, what difference would another 90,000 pounds make for a few days? Perhaps the DEC had a point, but it provided a sad commentary on the lake's ecological state.[113]

LCP had moved to the Syracuse area in 1979, when it bought out a chlor-alkali manufacturing facility that had previously been owned by Allied. The company had received a $4.5 million industrial development loan to encourage its siting in the area. In the period leading up to the request for discharges into Geddes Brook, the company had generated complaints from residents in the area about chlorine leaks.[114] Six months later, the problem of the discharges had not been resolved, to the point that the DEC was now threatening to shut the plant down. Mercury had become an especially noteworthy problem. Whereas the company had a permit to release .028 pounds per day into Onondaga Lake, it had consistently been dumping a pound. The state had uncovered over 100 violations of state water pollution control standards and was threatening a $1.07 million fine. The company was also being cited for worker safety violations by OSHA.[115] The results of the dumping were clear: The state had determined that the sediment in Geddes Brook was among the four highest mercury contamination sites in the state of New York.[116] The DEC soon upped the pressure on LCP by threatening criminal charges for its violations.[117] In what appeared to be a response to the DEC, the company laid off most of the plant's workers.[118]

The problems at the Metropolitan Wastewater Treatment Facility had still not been resolved. In fact, the county was under threat of a lawsuit from the Atlantic States Legal Foundation if it did not rectify the continuing pollution problems in Onondaga Lake related to the discharges of sewage. The county agreed to spend $30 million to fix a pump station in Liverpool and to upgrade the Ley Creek treatment facility.[119]

Attention now shifted to Oil City, an oil tank farm at the southern end of the lake. Oil had been leaking from the tanks onto the ground for years, and the DEC wanted to know how it was moving through the soil and whether it was contaminating groundwater and Onondaga Lake itself.[120] Seven months later, the scientists hired by the city to study the area's contamination were asking for more time. Given the legacy of pollution on the site, they apparently had only begun to scratch the surface. While the area had been an oil storage tanker facility since the seventies, it had previously been a center for salt drying beds, and then it had become a dumping ground for Solvay process wastes. It was estimated that over the decades, 250,000 gallons of oil had leaked into the area from various spills.[121] The city agreed to a two-month delay, and in March LCP pled guilty to criminal charges of polluting Onondaga Lake with mercury and paid $650,000 levied by a federal district court. By the end of April 1989, LCP had still not reopened, and the state was planning to add the site to its list of Superfund sites.[122]

That same spring, Senator Daniel Patrick Moynihan was in Syracuse at a meeting of the Common Council to push forward his proposal for $100 million plan to clean up Onondaga Lake. He accused Allied Chemical of "raping" the lake, and compared Allied's actions to those of Exxon, under fire for the Exxon-Valdez oil spill. Moynihan said, "Exxon did it in a day. Allied dumped a quarter of a million pounds of mercury in the bottom of that lake . . . They poisoned our lake for a century . . . We should never have let them get out of town . . . We should have locked them up." In response a company spokesman said, "We've always operated within the guidelines of regulatory agencies. We've been shut down since June 1988, and we've maintained a professional staff on site. The senator is not correct."[123]

In spite of all of this, the Carousel Corporation began to move forward with plans to build a large mall on the site where the Oil City facilities had been located. The mall would "be built on a 2 ½ foot thick slab of concrete so huge it'll 'float' on 250 feet of soft soil, sand, and gelatinous chemical waste left behind by Allied Corp.'s Solvay Process."[124]

It was also becoming clear that the deal that was made to pipe Solvay process wastes into the Metropolitan Sewage Treatment facility was a ruse, done in order to prevent the company from leaving town. Flocke, an engineer for the state DEC, stated to the *Syracuse Herald-American* what, in retrospect, should have appeared to be obvious: "It was sold to the people as co-treatment so Solvay Process would meet regulations by piping their wastes at Metro . . . It took 10 years to build, and after it was in operation, the 30-inch diameter pipes used at the plant were reduced to only 14 inches holes because the calcium and phosphates reacted to form a solid that slowly choked the pipe openings. More than 90% of the calcium went through Metro and was never treated. It was a hoax."[125]

A year later, Moynihan would still be pushing his "rape" analogy and continuing his support for $100 million in cleanup money for the lake. He seemed to be getting support for his hard-edged approach to Allied Chemical when State Attorney General Robert Abrams initiated a legal process that would require Allied to pay for the cleanup.[126] By July, the Senate Environment and Public Works Committee voted to allocate $5 million in seed money to get the process moving forward.[127] Later that year, an experiment to put sewage sludge over the old Camillus wastebeds was halted after a few weeks, due to concerns of residents about the odors. The project, by the company Bio-Gro, was designed to cut the costs of hauling the sludge to a landfill near Buffalo, while also turning the property into a "flourishing grassland and forest." But the process generated excessive ammonia that the company was having a difficult time controlling.[128]

On January 15, 1988, the Atlantic States Legal Foundation filed a lawsuit against Onondaga County alleging violations of the Clean Water Act and the New York State Environmental Conservation Law. The

New York State Department of Environmental Conservation joined the lawsuit, which was aimed at the Metropolitan Treatment Plant, and the intention of which was to bring it, and various sewage outflow points (in watershed tributaries), into compliance with water quality standards related to ammonia and phosphorus. The disposition of this lawsuit, more than twenty years in the future, would have important ramifications for sewage treatment in the Onondaga watershed, would indirectly lead to valid claims of environmental injustice, and would transform the county into a national model for green approaches to rainwater runoff management.[129]

On June 27, 1989, the State of New York Department of Environmental Conservation brought a civil lawsuit against Allied-Signal, for violations of the Comprehensive Environmental Response, Compensation, and Liability Act (a.k.a., Superfund) and various violations of state law. Under Superfund, legacy pollution from an industrial facility is the responsibility of that corporate entity, even if it no longer operates at the contamination site. This lawsuit would mark the beginning of a decades-long process requiring Allied to remediate the damage that it had inflicted on Onondaga Lake and the surrounding environs.[130]

Environmental organizing around Onondaga Lake evolved in the 1970s and 80s, reflecting broader trends across the United States (and globally), but with its own unique characteristics. Middle and upper-middle-class suburbanites were often at the forefront of resistance to the Allied's push for expansion. Such activists may or may not have had broader environmental commitments. But they were concerned with the immediacy of the threat that plant expansion posed to their quality of life, their health and safety, and likely their property values as well. But their mobilizing constitutes one part of a broader mosaic. The State of New York College of Environmental Sciences and Forestry provided a deep pool of researchers who could and did study the lake. Activists such as Steve Effler acquired their training at ESF. A supportive role for lake cleanup was provided by area fisherman, recreationalists, and members of the business community who saw eventual opportunities for tourism, and perhaps even real estate development. State and local governments were also exercising new regulatory authorities to put the squeeze on Allied, other industrial operations such as Crucible Steel, and Onondaga County.

Together these various forces placed Allied in an increasingly defensive position. Its shuttering was not likely a response to pressures from environmental groups, but when it did close, those most directly affected were the workers and the small business owners in the community of Solvay. As with other instances of deindustrialization, whether in Buffalo, Pittsburgh, or Detroit, the cessation of industrial production produced tremendous economic dislocations, but it also provided the opportunity for environmental revival, with potential economic benefits. If, however, anyone were to naively assume that with Allied's disappearance from the scene, ecological health would be returned to the lake in the near term, they would be severely disappointed. Allied's closing would mark only the beginning of a very long and very contentious process of litigation, negotiation, and plans for remediation.

PART III

Environmental Restorations

5

A Third Wave of Environmentalism

The Bumpy Path toward Restoration

Restoration efforts began at Onondaga Lake in the 1990s. The Onondaga Lake case is unusual in the sense that the principal agents of its contamination agreed, although under pressure provided by legal compulsion, to address their contributions. The two main culprits were Allied Chemical and Onondaga County. While this may suggest a resolution to this complex and long-running history, that turns out to be not entirely true. The restoration efforts, in fact, spawned new conflicts. This should not be surprising, however, because the very concept of restoration is the subject of intense debate within environmental circles, and, while its technical elements are often worked out by engineers and environmental scientists, its implications run deeply into fundamental questions about the relationship of humans to the natural world, questions that are the domain of social scientists, philosophers, and policy makers.

Environmental restoration could be seen as a third wave of environmentalism. The first wave in the U.S. focused on the conservation of natural areas, and has its roots in the nineteenth century, as Americans began to recognize that what was once believed to be a nearly infinite wilderness was being destroyed at an alarming rate. The second wave of environmentalism, which emerged in the second half of the twentieth century, focused on the impacts of industrial pollution, and as noted previously, evolved in reaction to local and global environmental contamination. Rachel Carson is an important figure in this movement, but she was not the first, and her writing can be seen as continuous with what came before, not a radical break from it. The environmental justice movement, with its focus on urban deterioration and toxic disposal, grew out of these earlier movements, and marked a significant advance in thinking through them in terms of racial and economic equity.

Environmental restoration, or restoration ecology (I use these terms synonymously here), pushes the concept of conservation a step further. Given damage done to so many environmental systems in the U.S. and globally, it is insufficient to simply protect what is left. Rather, we need to bring back, or restore, when possible, that which has been lost. Aldo Leopold was among the first to write about and practice of environmental restoration in the U.S., although in a broader sense ecological restoration has been part of the human experience for thousands of years. William Jordan and George Lubick, for example, propose that early humans' use of fire to manage habitat was a form of ecological restoration, as were certain agricultural practices.[1] Leopold, for his part, attempted to restore parts of the prairie near his home in Wisconsin. Leopold, in turn, inspired a younger generation, including Jordan, who is considered one of the first to think deeply about, write about, and practice as a life's work, environmental restoration. He, like Leopold, worked at the University of Wisconsin Arboretum in Madison.[2] Early restoration efforts, then, linked as they were to the conservation movement, focused primarily on natural areas, in the case of Jordan, on prairie grasses and forests. But, as with the broader environmental movement, the concepts developed in the field began to move into the more greatly damaged ecologies of urban and industrial areas.

With a shift of focus to industrial sites and toxic waste, the term "remediation" took hold. Remediation implies repair of a specific problem, but not necessarily revitalizing an entire system. To make a comparison to medicine, taking an antibiotic might cure someone's strep infection, but it doesn't, by itself, create a healthy equilibrium in a body with multiple problems. It won't cure high blood pressure or heart disease or a damaged liver or faulty eyesight. Even if the infection is staunched, in other words, it remains only one piece in a larger and complex homeostatic system. That doesn't diminish the power of antibiotics for the particular role that they play. It only points to their limitations. Making a severely damaged ecosystem healthy requires more than repairing its separately damaged parts.

Remediation is, then, a narrower term that tends to be applied to worst case environmental scenarios. According to the Department of Labor, "Environmental remediation is the removal of pollution or contaminants from water (both groundwater and surface water) and soil.

These waste products are removed for the protection of human health, as well as to restore the environment."[3] Very often, for example, remediation is associated with the cleanup of atomic waste sites[4] or urban brownfields.[5]

Given that remediation is a form or restoration, the terms are sometimes used interchangeably. Allison, for example, distinguishes two tracks for ecological restoration. One involves taking land that was "utterly destroyed," and which had little if any ecological function, and improving it to whatever degree possible. Here the goal is to produce "whatever kind of ecosystem was possible on that derelict land." A grassland might be turned into a marsh, a forest into usable agricultural land, a brownfield into an urban park.[6] The original system might not be replicated, but the restored area could still have greater ecological integrity.

Many, perhaps most, ecological restoration projects will fall somewhere between restoring a marginal landscape and remediating an utterly destroyed local system. The Onondaga Lake project has involved elements of both. It has involved reestablishing native species of flora and fauna while also remediating highly contaminated areas along the lake's shorelines and in the lake bottom. Willow trees planted in the Solvay calcium chloride beds did not replicate native species, but they were a practical way to stabilize the mounds. Bringing back Atlantic salmon and other species of fish that were eliminated when the lake was in its worst condition is an attempt to reestablish previously intact aspects of the system. Cleaning up the wastebeds and other highly contaminated areas is more like brownfield remediation.

Two lawsuits in the 1990s initiated the Onondaga Lake cleanup. The New York State Department of Environmental Conservation sued Allied Chemical, demanding cleanup of their contribution to the lake's contamination. At first the company resisted, but when it bought Honeywell and took Honeywell's name, in an apparent attempt to improve its reputation, it became more cooperative with the state. The state and Honeywell entered into negotiations and arrived at an agreement, the primary elements of which were dredging and capping. The total cost was estimated at one half billion dollars, the most expensive remediation effort in American history. During the same period, the Atlantic States Legal Foundation brought suit against Onondaga County to stop its dumping of sewage into the lake. Now the two main contamination problems, which had persisted for decades, would finally be addressed. But these lawsuits and eventual agreements marked only the beginning and not a predetermined or inevitable restorative outcome.

Capping and dredging did not, by itself, mitigate the mercury contamination of Onondaga Lake. It took another intervention, spurred by the work of Steve Effler and others at the Freshwater Institute, who discovered that introducing nitrates into the lake bottom would reduce methylmercury by a series of chemical processes. Moreover, the dredged material would have to be deposited somewhere. A site of a former wastebed in Camillus became the choice. When a wastebed had been located there earlier in the twentieth century, few people lived in the area. Beginning in the 1960s, it became the site of a prospering suburban community. Residents experienced terrible smells and became understandably concerned with the impacts of the off gasses on the health of themselves and their families. The Department of Environmental Conservation minimized any concerns, but opponents were undeterred and brought lawsuits against the state.

Members of the Onondaga Nation were displeased with the remediation plan as well. It did not, in their view, "restore" the lake. One shoreline was replaced with a retaining wall, reducing the lake's size. At one point, some of the capping materials, mostly sand and gravel, slipped from their original location, exposing the toxic mud underneath. Representatives of the Onondagas questioned the mercury science of the Freshwater Institute and ESF. Most importantly, they never had a seat at the table as negotiations were being conducted on the fate of the lake. Their exclusion reinforced a sense of marginalization in terms of participation in land use in Central New York, and the entire region, that had been practiced for decades, if not centuries.[7]

The federal district court case, brought by the Atlantic States Legal Foundation, resulted in a directive to build several satellite sewage facilities, one in an historically Black neighborhood in South Syracuse. It would involve construction of a chlorination facility and the demolition of a number of houses. The community would be disrupted and housing values would be negatively affected. Opposition to the plan grew beyond the neighborhood, including from the law firm that originally brought the lawsuit. When Syracuse refused to deed the land to the county to construct the facility, the county went to court and won. To avoid further expansion of the plant, however, and to avoid placing another one in a lively downtown commercial district, activists and a new county executive convinced the court to accept another option to reduce the inflow of water into the main sewage plant during high rain and snow-melt events. Save the Rain became a national model of how to deal with excess runoff in combined runoff and sewer systems. It was embraced by public officials, environmentalists, representatives of the Onondaga Nation, and citizens. It also saved money.

Efforts at Onondaga Lake have involved expensive and extensive re-mediation. But it would be difficult to contend that Onondaga Lake has been "restored." It has been improved. The capping, dredging, and in-jecting has lowered methylmercury levels in the ambient water. Much of the lake is graded satisfactory for swimming and fishing, and many fish species thrive in the lake. Marshes have been restored. Trees and meadows now exist on parts of the lake's shoreline once dominated by exposed wastebeds. A pedestrian and cycling path around the lake is called "Restoration Way."

For the foreseeable future Onondaga Lake restoration efforts will re-quire ongoing technical interventions. Nitrates are being continuously added to the lake bed to keep methylmercury levels under control. How long this will have to occur is, as will be seen, a matter of dispute. What-ever the case, groundwater will have to be continuously filtered to keep runoff from old wastebeds recontaminating the lake. The remnants of Oil City will have to be kept sequestered in a plastic casing to prevent its contents from leaching into the groundwater and the lake itself. The cap on the lake bottom will have to be monitored and maintained to prevent it from subsiding and allowing the contaminants beneath it from enter-ing the lake's waters. How long this will be the case is not entirely clear. In other words, the lake, at this point, would not meet historical bench-marks that preceded establishment of Allied Chemical. In that sense, it does not meet the standards of restoration set forth by many scholars and practitioners in the field.

The emphasis of restoration projects has been toward making the lake useful for human activities. Various studies that have been conducted on the lake's ecological status have focused on the scientific data, while also making recommendations regarding the recreational opportunities available. They address whether the lake is safe to swim in and whether fish in it can be safely eaten. As the lake's contamination declines, then, it seems likely that more humans will be drawn to the lake, and it will become a focus for various forms of public and private economic development. A bicycle path has been created, an amphitheater has been constructed, and plans for a beach have been completed. Economic development such as this generates its own imperatives. If often goes hand-in-hand with forms of gentrification. It is not clear whether lakefront homes and condos will start to crowd out natural spaces, but, if they do, they may negatively impact steps made toward restoring the lake and the natural systems that border it. Onondaga Lake is somewhat unique in that most of the land surrounding it is owned, or at least controlled, by Onondaga County. That provides a buffer against untrammeled economic development, but not a guarantee.

An interesting parallel exists with the infamous Gowanus Canal, in Brooklyn, New York. The canal was one of the most contaminated waterways in New York City. It was dirty, smelly, and often filled with dead fish and even the occasional dead body. In the early 2000s, however, as an explosion of development projects cascaded across New York City, the Gowanus neighborhood became a target for large residential development projects. Local residents resented the intrusions and mobilized to keep them out, with some success. When the U.S. EPA stepped in to consider labeling the canal a Superfund site, anti-development forces saw this as an opportunity to frighten off developers, who would be less likely to invest in an area that was on the national registry of toxic contamination sites. They supported the designation. One resident developed a tagline for their local group with the phrase, "Gowanus Canal: Superfund Me!" in large block letters.[8] When the EPA affirmed the canal's toxic status, developers pulled out.

I am not suggesting here that the Superfund designation for Onondaga Lake, which is still in place, should remain there to protect against overdevelopment. I'm suggesting that environmental restorations can have unintended and not fully recognized consequences that might, at some point, undermine the extensive and determined efforts that brought about those restoration efforts in the first place.

Given the complications of restoration, in terms of execution and outcomes, democratic participation in it should be a high priority. In Eric Higgs's words, "The value of a restoration project can be measured in part by its contribution to democratic participation, and the greater the participation the greater the value."[9] In the case of Onondaga Lake, the stakeholders with the most say were a very large global corporation and the state's Department of Environmental Conservation. It would seem only fair that the voices of the Indigenous people whose cultural traditions were most strongly tied to the lake going back for millennia, would have at least an equal voice at that table, but they didn't. Moreover, they should be included in future decisions about the lake's future to a greater extent than currently appears to be the case.

Over three decades, a consensus emerged, through fits and starts, that Onondaga Lake was worthy of restoration. The impulse for some kind of restoration intensified with the shuttering of Allied's facilities in Solvay. As steps were proposed to move forward, recognition emerged of how immense and complex the project of a lake cleanup would be. The question of what that restoration would involve had not yet been determined, or even imagined in any practical terms. Still, the early 1990s marked a period of optimism. After all, Allied was no longer operating, so its contaminants would no longer be cast into the lake, its tributaries, or the surrounding landscapes. Congress had agreed to provide funds of up to $1.5 million (a woefully inadequate sum in retrospect). The Carousel Mall and Franklin Square, two downtown redevelopment projects, had further spurred interest in the lake's restoration. At a meeting in a barn at the Old Barge Canal terminal, Mayor Tom Young asked, "Does it make any sense to have the kind of land use we are dreaming of, if we don't have a clean lake?" A restored lake was couched as essential to downtown revitalization.[10] For decades, the economic arguments tended to favor industry; now they were shifting to environmental protection. For the first time since the late nineteenth century, the lake itself was perceived as a potential economic asset. But while the economic arguments gained momentum, they were inadequate to provide for the quality of lake restoration envisioned by members of the Onondaga Nation. That clash would eventually manifest in very specific policy terms.

Inching toward Remediation

On Earth Day 1990, the *Syracuse Herald American* announced that Syracuse was a "winner in the clean air fight." The article celebrated how much cleaner the air had become since the days when pollution obscured the view of Onondaga Lake from the hills surrounding it. While improvements in automobile emissions accounted for some of the difference, much of it was the result of the closing of the Solvay Plant and Crucible Steel. However, this wasn't an unqualified victory for the environmental movement. Improvements in air quality had resulted as much from declines in industrial production as from the imposition of environmental standards. Syracuse represented a microcosm of the industrial Northeast and Midwest in this regard.[11] This was true in Pittsburgh, Buffalo, Cleveland, and Allentown, Pennsylvania, among other cities. But while declines in industry may have had a salutary effect on air pollution, it did not remove the massive amounts of toxics that had been casually dumped into those industrial landscapes for decades. To revive them, for economic development, much less for ecological integrity, would take time and considerable economic investment.

Carousel Mall was now being constructed on a toxic waste site in an area that had once been a landfill, and upon which had been deposited thousands of tons of waste from the Solvay Plant. As a result, the area had to be dredged, and cleansed, and an elaborate system of pumps installed to keep out groundwater and to prevent polluted water from inundating the area from the lake. From a health and environmental prospective, it seems an odd place to construct a mall, but the land was cheap, and as a form of brownfield development, it garnered New York State grants.[12] Moreover, located at the southern tip of Onondaga Lake, the mall provided a symbol of the economic possibilities of the lake's renewal. However, the complicated and costly engineering and construction involved was a harbinger of later cleanup efforts. Would it be possible to return Onondaga Lake to its "natural" state, given the amount and variety of contamination sources? Over time, it became clear that the area could only be protected via a series of elaborate technical interventions which, if the lake were to avoid continuing contamination problems, would need to be sustained over time, although how long is a matter of dispute.

Members of the fishing community had long been advocates for lake restoration, and they were also hopeful about its prospects. Despite, its various ecological stresses, the lake was teeming with aquatic life. In June 1991, an article in *The Post Standard* listed the many species of fish that lived in the lake, including whitefish, bluegills, sunfish, carp, and even walleyes. The size, number, and variety of fish were undoubtedly connected to that fact that so few people were fishing for them. The reason was obvious: High levels of mercury were contaminating all the fish, although some more than others. Carp and other larger fish were clearly off-limits, but sunfish and other panfish, because of their short lifespan, absorbed fewer toxins, so there was hope that at some point in the future, they might be edible, at least in small amounts.[13]

Unfortunately, Allied's closure did not terminate the practice of draining industrial effluents into the lake. The Niagara Mohawk power company was discharging salty water into Onondaga Creek as late as 1991 and had apparently been doing it for years. Company wells were the source of the problem. The DEC had ordered the company to stop and threatened fines of $10,000 a day for discharges before 1987, and $25,000 a day for discharges after 1987. But the discharges did not stop, and they were only made public when two Upstate Freshwater scientists tried to understand why Onondaga Creek salt levels were above normal. Their investigation led them to the discharge pipes.[14] In the DEC's defense, it is possible that it failed to effectively monitor the salt discharges because they were disguised by the high levels of chloride already in the lake from Allied's operations. Since the plant's closure, those levels had dropped, making other point sources more visible. But why the Niagara Mohawk discharges were salty remained something of a mystery. Upstate Freshwater suggested that they may have been running across old Allied brine.[15] This would become one small example of how complex and intertwined the contamination problems would turn out to be.

The finding of salty discharges from the Niagara Mohawk facility led to renewed interest into whether Onondaga Lake was naturally salty. An editorial in the *Post-Standard* suggested this notion drew on observations from "early Jesuits" in the seventeenth century and the remarks of a Syracuse University professor of engineering, who noted the lake's saltiness in 1968.[16] But Timothy Mulvey, executive director of the Onondaga Lake Management Conference, disagreed. Salt springs did discharge into Onondaga Creek, and these were attested to by Father LeMoyne when he entered the area in the 1645. Salt became the basis for much economic activity. More than a hundred salt manufacturers operated near the lake by 1826 and dumped their waste into it. But the natural salt deposits were located at the southern end of the lake. Given the depletion of the salt beds, naturally occurring salt was no longer being discharged into the lake from there. The source, Mulvey concluded, must be the infamous Solvay wastebeds. There had to be some outside source, because Onondaga, as a whole, was never a naturally salty lake.[17]

In 1991, a mysterious "goo" was found at the bottom of the lake. Scuba diving scientists had discovered the substance while attempting to study the lake's plant life. At first, it wasn't even clear whether this was some new contamination problem or whether it had been there for a long time and covered up by sediment. Lee Flocke, a regional water quality engineer, didn't seem overly concerned. "The stuff had been there for so such a long time that a few more weeks probably won't cause a problem with the lake." But DEC chemical analysts eventually undermined Flocke's sanguine assertion. The goo turned out to be a toxic stew composed of benzene, chlorobenzene, napthlalene, fluorene, and mercury. While officials were not immediately willing to lay the blame at the feet of Allied, given the multiple sources of pollution that had historically been in the area, it seemed like the most likely source.[18] How to handle this goo became one of the key points of contention with regard to lake restoration efforts.

Meanwhile, a 1992 Senate hearing of the Committee on Environment and Public Works, chaired by Senator Daniel Moynihan, focused on proposals to divert outflows from the Metro sewage operations into the Seneca River. The proposal had the support of Timothy Mulvey, Steve Effler, and Senator Moynihan. There was some disagreement among witnesses as to whether full diversion, or a plan for partial diversion, would be the best option. Effler and County Executive Nick Pirro favored the partial diversion, due to potentially negative impacts full diversion could have on the Seneca River and due to cost factors.[19] One witness, Raymond Oglesby of Cornell University, suggested that if the diversion was completed, it would solve Onondaga Lake contamination issues. The mercury in the lake, he contended, was being locked up in the sediments, and, therefore, did not pose a long-term problem. He opposed dredging, because he believed the disturbance of the sediment would release contaminants and potentially do more harm than good.[20] In the end, the proposals went nowhere, primarily, it seems, due to a lack of adequate state and federal matching funds.

By 1994, after decades of various attempts to mitigate sewage discharges into Onondaga Lake, a tertiary level sewage treatment plant had still not been constructed. Edward Kochman, Deputy County Executive for Onondaga County, was questioning the value of such a facility. "Even if the county improved the sewage treatment plant so that no raw sewage spilled into the lake, there'd still be mercury on the bottom and chemical waste on the shore. The clarity of the water would only be a few feet and the optimal amount of oxygen would exist for only a few more weeks each year." The cost of $1300 a year for ten years for each taxpayer, he suggested, would simply not be worth it.[21] He resurrected the proposal to reroute the sewage directly into the Seneca River, which would involve the construction of a seven-mile-long pipe, and where the unstated implication was, that the pipe would provide a direct route for minimally treated sewage into Lake Ontario. But the proposal again failed to gain traction.

In a *Post Standard* article, Timothy Mulvey recounted the long and troubled history of attempts to stop sewage discharges into Onondaga Lake. In 1954, the Onondaga County Board approved a plan to build the Metro sewage systems on the lake. Part of the proposal involved Kochman's solution, rerouting the sewage to the Seneca River at a cost of $3.6 million, but that part of the plan was never completed. In January 1966, the county withdrew the plan, because it was believed that the dumping of effluents from the sewage plant would dilute the industrial waste from the Solvay Plant. In 1979, a new Metro plant was finished, which authorities claimed would end the dumping of raw sewage into the lake, but it failed to live up to that promise. When Solvay closed down in 1986, the sewage facility became the single largest polluter in the lake. In 1988, the Atlantic States Legal Foundation and the New York Department of Environmental Conservation sued to end the sewage discharge into the lake. As part of the settlement, an analysis of the lake's pollution problems was conducted that concluded with a finding that the only way to resolve the problem was to reroute the sewage around the lake and into the Seneca River. Another study by the Army Corps of Engineers arrived at the same conclusion. Estimates for a new sewage treatment plant and a diversion pipe were now in the $200 to $300 million-dollar range.[22]

A satirical letter to the *Herald-Journal* in 1994 suggested that Onondaga Lake could be a tourist destination, the centerpiece of a toxics tour of the Syracuse area. "While we wait around for the bureaucrats to begin the cleanup," the author suggested, "let us not waste this money-making potential. Let's use the pollution to bring in tourist dollars, provide jobs for people and invigorate the local economy. Don't think of pollution as the problem—Think of it as the solution."[23] While the article was obviously written tongue-in-cheek, there is actually precedent for using toxic sites to attract tourists. Butte, Montana, is a primary example. The Berkeley pit is massive, stunning, and strangely captivating in its ecological devastation. At an even more extreme level, the still highly contaminated Chernobyl nuclear facility had become a tourist destination (until, of course, the Russian invasion in 2022). From a purely pecuniary perspective, however, Onondaga was not so attractive. After all, it looked relatively normal from a distance. The only indication of its massive contamination problems at the time were the awful smells emanating from it, which would probably not be a major draw for tourists.

Ten years after the closing of the Solvay plant, there had still been no activity on the old site. A number of small companies had located on the former grounds, drawn in partly by two cogeneration facilities that were producing relatively cheap electricity. Tax breaks had also been included as an incentive. So, although 500 jobs had been created, there was still a gaping hole in the city's tax base. As a result, the provision of public services posed a particularly difficult challenge.[24]

In October 1996, the Onondaga Lake Management Conference, a consortium of local, state, and federal officials charged with improving the lake's water quality, approved a restoration plan of $3.5 million. $1 million would be spent on restocking fish in the lake, and $2.5 million on various cleanup efforts. $100,000 was approved to study how to stabilize the notorious Solvay wastebeds, which had been given the nickname the "White Cliffs of Dover."[25]

Promises of an Onondaga Lake cleanup featured prominently in the 1997 Syracuse mayoral race. Both campaigns kicked off on the Onondaga Lake shore. While such symbolic actions were unlikely to provide additional progress in tangible cleanup efforts, they were an indication that the lake had become a potent political symbol, which candidates and elected officials could ignore at their peril.[26]

Finally, in October 1997, the Onondaga County Legislature approved a plan for a tertiary sewage treatment by a wide margin and in doing so, resolved a long-running legal battle with Atlantic States Legal Foundation. Rather than reroute the sewage to the Seneca River, as had been proposed by some, the county agreed to improve the Metro sewage plant. It would stop the flow of raw sewage by 1999, remove ammonia from the discharges by 2005, and remove phosphorous by 2006. The result, if the plan were to be successfully implemented, would significantly reduce sewage discharges into the lake. The total costs were estimated at close to a half billion dollars. The county's share would be $120 million dollars. The state would contribute $260 million, and the federal government would contribute $100 million. A typical family's sewage fees would rise from $227 a year to $387, manageable for middle-class families, but not an insignificant amount.[27] A version of this plan would eventually be adopted and implemented.

By 1999 it had become clear that an increasingly diverse fish population was living, if not thriving, in Onondaga Lake. A 1960 survey revealed that virtually all of the lake's fish were carp. A 1990 survey found thirty different species. When Syracuse Professor Neil Ringler undertook a survey in 1999, he found fifty-two fish species. The return of a more diverse population had already been obvious to local anglers. A bass club was now running a tournament on the lake, and bass as large as four pounds were being caught.[28] The County Health Department advised loosened restrictions on fish consumption, based upon slight declines in the mercury contamination of some fish. However, some researchers were skeptical as to whether this was warranted. While some species, such a small mouth bass, showed mercury contamination below EPA-sanctioned levels, walleye concentrations, at 1.9 parts per million, were twice the accepted standard. And, while it might be expected that mercury concentrations in the lake had fallen, given the closing of Allied's facility, evidence began to emerge that these expectations were not being met. Charles Driscoll, a Syracuse University expert on mercury pollution, conducted an independent analysis, concluding that mercury levels in the lake were virtually unchanged since the 1970s. Some suggested that Allied had pressured the county into changing its advisory, but county officials denied these charges.[29]

A Corporate Rebranding

Allied had been resisting attempts to have it pay for the lake's cleanup after it left the area in 1986. It had even sued to require Onondaga County to take responsibility for its share of the pollution, on the grounds that Allied had been encouraged by the county to unload its wastes into the lake in order to control phosphate levels from the county's sewage. (In fact, there was some truth to the charge.) Still, Allied intransigence was also evidenced in its ignoring of a court order to conduct studies of the lake. The company was fined for missing deadlines in 1997 and 1998, but things changed dramatically in 1999. AlliedSignal sought to purchase Honeywell, a company associated with heating, cooling, and aerospace, allowing it to adopt the name and brand of the smaller company, and to shed its historic association with chemical manufacturing and toxic waste. Or, as the Honeywell website states it, "in 1999, Honeywell was acquired by AlliedSignal, who elected to retain the Honeywell name for its brand recognition."[30] To satisfy regulatory authorities, the companies would have to resolve nagging legal problems, including the Onondaga Lake cleanup issue. Honeywell's remediation director conceded, "If you are a chemical company that has been in operation since the 19th century, that is part of your legacy. The important thing is that we are going to take responsibility for that."[31] This seemed to mark a monumental shift away from what had been AlliedSignal's intransigence.

In March 1999, a private company from New Jersey proposed buying the Onondaga County sewage system, which included the Onondaga Lake facility and eight others, claiming that it would make the improvements required by the $380 million cleanup project. County officials expressed some skepticism about the offer. The private firm claimed that it could run the operation on a more efficient basis, but the possibility of turning a profit on a system with such a troubled history seemed unlikely. In spite of the county's skepticism, Upstate Fresh Water Institute, a local environmental group interested in the lake's cleanup, endorsed the company's plan. Syracuse University Professor and Board Member Robert Hennigan stated, "I like their conceptual approach to the whole thing. It just makes sense." The company was proposing to build a $100 million new sewage facility on the Solvay wastebeds,[32] but the transfer never took place.

In June 1999, the Onondaga Nation entered the picture, filing a petition with the U.S. Environmental Protection Agency under President Clinton's 1994 Environmental Justice initiative, which focused attention on environmental issues in underserved and minority communities. The claim was aimed at the Tully gravel mine at the headwaters of Onondaga Creek. The suit charged that water quality had been damaged, fish population undermined, and Native American archeological sites destroyed. Plaintiffs demanded that the mine immediately cease operations.[33]

Meanwhile, the Honeywell/AlliedSignal merger seemed to signal a new era of cooperation with local and state officials. As a start, the company agreed to demolish the last remaining structures from the old Solvay works and to remediate the remaining toxic substances. Among the eight buildings to be demolished was a chlor-alkali manufacturing facility, originally owned by Allied, but sold to LCP chemicals in 1979.[34]

Honeywell also began removing samples from the bottom of the lake, in the first step of their proposed cleanup process. Skeptics, however, wondered whether test results would be valid, given the company's obvious interest in the findings. Both environmentalists and public officials called for a panel of independent scientists to review Honeywell's findings.[35]

Another potential merger raised questions about continuing corporate commitment to a cleanup. General Electric, a company with a decidedly mixed environmental record, was proposing to merge with Honeywell. GE had mounted a legal challenge against the national Superfund law, questioning its constitutionality. The company had become notorious for its years of dumping PCBs into the Hudson River near Schenectady. It was also resisting demands for a massive dredging operation there, estimated to cost $1 billion. The Superfund challenge was an obvious attempt to undermine any legal authority requiring the company to foot the bill for remediation. Onondaga Lake advocates were understandably concerned, given GE's unwillingness to take responsibility for the Hudson River contamination, that a merger might have negative implications for the lake.[36] But when the deal collapsed due to European Union objections, Honeywell's previous position, which was at least publicly supportive of a cleanup, remained unchanged.

Still, by Fall 2001, local officials and environmentalists were expressing impatience at the lack of progress being made by Honeywell to follow through with its cleanup plans. The New York State Department of Environmental Conservation had rejected the study done by Honeywell of all contamination sites, on the grounds that it was incomplete. A spokesman for the company denied that this was major setback but instead was a fairly standard part of the process when dealing with complex toxic legacies. The county's environmental director, David Coburn, was less optimistic. He stated that rejection of the study pointed to larger problems with compliance. "Honeywell," he said, with some exasperation, "has to be more genuine in the process. It seems like every time you turn over another rock on the old Allied property, you find more contamination. The problem is their contamination is so widespread and so pervasive, you get the impression only time and nature will take care of the whole problem."[37] The comment would prove to be prescient.

In spite of the problems with the site survey, Honeywell was proposing some substantive steps forward. Geddes Brook, which was contaminated with mercury from an old culvert pipe, would be dredged. A more ambitious proposal involved constructing an underground wall to seal off what was left of the now demolished LCP Chemical plant in order to prevent remaining toxics from leaching into the lake. Honeywell also proposed a system to collect polluted groundwater from the Willis Avenue petrochemical plant and to remove seventeen tons of the tar-like substance that had been created when the facility was burned and its contents unceremoniously dropped into five lagoons. Known formally as the Semet Residual Ponds, a more colloquial name for them, the "stinking tar beds," offered a more accurate description. But whether the materials could be converted into usable products, such as light oil and driveway sealant, was open to question. The company also claimed that it was working on a plan to deal with the mountainous Solvay wastebeds and that it was committed to completing a survey of the lake bottom itself.[38]

As various surveys and remediation plans to deal with Solvay sites were slowly moving forward, expansion of the Carousel Mall in Syracuse was inching uncomfortably close to major toxic waste sites at the southern tip of the lake. When ground was broken for a new hotel, county officials expressed mild alarm about the potential for unleashing previously contained materials. When the mall was first built, contaminated soil containing chemicals for dry cleaning, including trichloroethylene, toluene, acetone, and vinyl chloride, were removed from the site and sequestered into a cell, contained by heavy plastic sheets, and buried under the mall parking lot. Dick Brazell, a regional official for the state DEC, said, "They know they can't puncture that cell. It's something they can't touch." The state was taking the position that no contaminated soil be permitted to leave the site. Either it would have to be remediated on site or left alone.[39]

In June 2002, while work seemed to be stalled at the former Solvay site, the county was moving ahead with a major expansion of the Metropolitan Sewage Treatment complex. The largest piece and most complex part of the project involved construction of the main treatment plant at a cost of $125 million. When completed, it would have the capacity to process large quantities of phosphorous and ammonia, keeping these out of Onondaga Lake. Other elements included improvements to the Kirkpatrick Street pump station, expansion of a storage system to prevent overflows into the lake, and floatable controls facilities, designed to prevent "floatable" wastes from entering creek or lake waters.[40]

In August 2002, nearly a year after New York State rejected the study that Honeywell had submitted as part of its proposed cleanup effort, elements of it went public, and the reason for the state's rejection became clear. Honeywell claimed that half of the mercury in Onondaga Lake resulted from the Metropolitan Sewage Plant, and thus the county was responsible for it. On the face of things, this seemed to wildly overstate the county's responsibility, and, by the same token, completely understate AlliedSignal's contribution.[41] In general, sewage facilities do not discharge large quantities of mercury, because mercury isn't present in the sewage itself or included in the treatment process. The only source of mercury discharged from the Syracuse plant would be its processing of Allied/Signal or LDP effluents.

Legacies of Toxic Contamination

Perhaps the company had been testing the DEC to see whether it could get away with minimizing its contributions. A second version, released a year later, was more complete. It involved 6,000 samples, some collected by Honeywell, but also drawing on those collected by the New York Department of Environmental Conservation over an eight-year period beginning in 1992. Water, soils, sediments, and organisms were all tested.[42] The findings were sobering. The Remedial Investigation (RI), as it was called, examined fifteen sites, each with its own legacies of contamination. The main Solvay plant produced soda ash and related products. It was the main source of Solvay wastes that covered the area. It included a Petroleum Storage Area, and chlorobenzene hot spots associated with the Willis Avenue plant. Together, they generated a virtual witches brew of toxics that either leaked or were simply dumped into the surrounding tributaries, groundwater, and the lake itself. The area abutting Willis Avenue, just north of the main plant in the town of Geddes, was where chlor-alkali product and chlorinated benzene were produced. Contamination there included many toxic chlorinated organic compounds in addition to mercury. Allied discharged wastes from this plant into the East Flume, operating from 1918 to 1977, utilizing the diaphragm and mercury cell process.

The East Flume was a canal excavated in 1918 expressly to carry wastes from the Main Plant and Willis Avenue plant into Onondaga Lake. Stormwater from the surrounding areas, including the Village of Solvay itself, was diverted into the flume. It became one of the primary paths of mercury into the lake, and, given its high levels of contamination, it continued to contaminate the lake.[43]

What's known as the Willis Avenue Ball Field Site was a dumping ground, according to the New York State DEC and USEPA, for "Allen-Moore diaphragm cells, and related graphite, laboratory vials and flasks, construction and demolition debris, miscellaneous metal debris, and boiler slag." In other words, it was an all-purpose disposal site. In the 1960s, the area was turned into a baseball field.[44]

Tributary 5A into the East Flume "originates from a culvert north of the railroad tracks on the west side of Willis Avenue. It runs near the Semet Residue ponds and eventually enters the lake. Crucible Steel was just to the north of it, so its facilities' wastes were dumped into it, which included heavy metals and mill scale wastes (a flaky surface of iron oxides found on milled steel)." Mercury, PAHs, BTEX, dioxins/furans, oils, and possibly PCBs had also been detected in the ponds.[45]

Wastebed B had been a disposal site for Solvay wastes from 1908 to 1926. Syracuse City sewage sludge was also piped there from 1959 to 1965, to mix with the Solvay waste to minimize the sewage smell. At one point, the East Flume was excavated, and dredged materials were

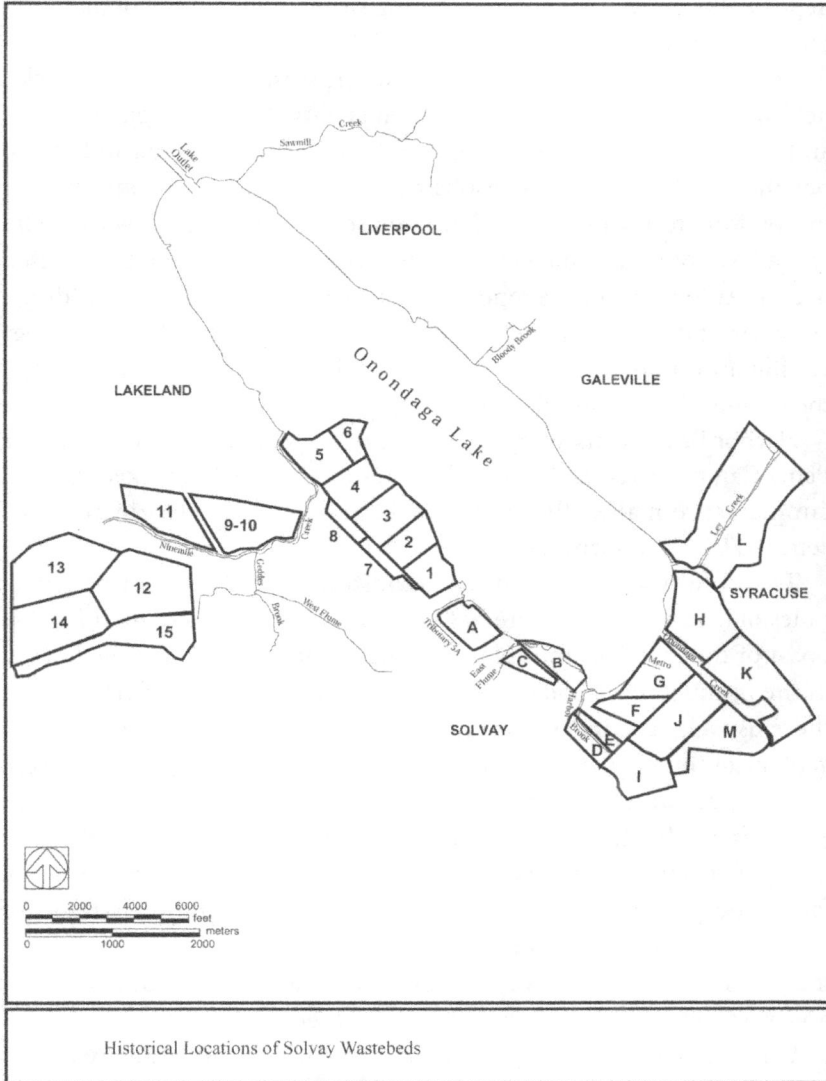

The map contains the following labels:

Creek
Sawmill Creek
Lake Outlet
LIVERPOOL
Bloody Brook
Onondaga Lake
LAKELAND
GALEVILLE
6
5
4
3
2
1
7
8
11
9-10
Nine mile Creek
Geddes Creek
13
12
14
15
West Flume
Tributary 5A
East Flume
Ley Creek
SYRACUSE
L
H
A
C
B
Metro
Harbor Brook
G
F
J
K
M
E
D
SOLVAY
I
Onondaga Creek

0   2000   4000   6000
feet
meters
0   1000   2000

Historical Locations of Solvay Wastebeds

Figure 5.1. Allied Wastebeds (2002). Courtesy of the Honeywell Onondaga Lake Cleanup website.

deposited there. It had been an ongoing site for the disposal of materials related to the flume's maintenance.[46]

The so-called Penn-Can property, to the west of the railroad tracks, held wastes from an asphalt production facility started by Semet-Solvay, and then operated by AlliedSignal until 1983. Seven hundred and fifty to one thousand cubic yards of asphalt-related materials were buried there in a pit forty feet wide, 165 feet long, and seven feet deep. A few feet of fill and a layer of permeable fabric known as a geotextile were then tossed over it. At the time of the report, the property "consist[ed] of buildings, aboveground storage tanks, and a gravel parking lot, with limited vegetation around the periphery."[47] Naphthalene continued to leak out of the pit into the surrounding areas.[48]

Harbor Brook runs along a CSX railroad bed, to the southwest of the Penn-Can property, and nearer the lakeshore. As such it picked up contaminants from all of those sites, including mercury, chlorinated benzene, BTEX, PAHs, and DNAPL.

The Solvay wastebeds themselves constituted 2000 acres, or 8.1 square miles of land along the shore of Onondaga Lake that were used for disposal of the massive amounts of calcium chloride that were generated in the production of soda ash. But other materials were often dumped there as well, as convenience dictated. The wastebeds hold a kind of geological timeline of the contamination process. As some places were filled, others were opened to replace them. Wastebeds 1–8 consisted of 315 acres of what had, at one time, been marshland. Allied used it for a dumpsite from 1926 to 1944. It was closed in 1944 after its dike famously failed and piles of mud and contaminants flooded into the streets of the Village of Solvay. As a result, Wastebeds 9–11 were opened, and remained so until 1968. Solvay wastes, brine purification sediments, and boiler water purification wastes ended up there.

Wastebeds 12–15 had a particularly noxious history. They operated from the 1940s until 1986. They received brine purification sediments, mercury cell wastewater, boiler water purification wastes, and boiler slag and fly ash. In 1986, the Metro sewage plant deposited liquid and solid sludge there. From 1981 to 1986 the wastebeds were used to send overflow water from Metro as a "chemical reagent for phosphorous precipitation" which was then funneled into the lake resulting an additional source of Solvay wastes.[49]

The report's summary was grim:

Due to the extensive volume of waste material (over 90 million cubic meters) in Wastebeds 1 to 15 alone, it is anticipated that the load or discharge
of contaminants to Onondaga Lake will continue into the foreseeable future. To some extent, discharges from the wastebeds are somewhat limited due to the low permeability of the waste material. Nonetheless, the
sheer volume of the wastebeds is sufficient to yield a substantive source
of ionic constituents, and in some case, organic compounds and mercury
to the lake.[50]

A potpourri of toxic organic compounds permeated the entire area,
including benzene, toluene, ethylbenzene, xylenes, chlorinated benzenes, polycyclic aromatic hydrocarbons (PAHs), PCBs, dioxins, and
furans. Solvay wastes were mostly calcium chloride, calcium silicate, and
magnesium hydroxide, but also included calcium carbonate, aluminum,
iron oxides, calcium hydroxide, calcium sulfate, ammonia and arsenic,
copper, lead, nickel, and zinc. These wastes entered all of the creeks that
ran through the property: the West Flume, Geddes Brook, Nine Mile
Creek, and the East Flume.[51] The report estimated that three million
cubic meters of waste were flushed into and then out of the East Flume
alone.[52] Moreover, "Based on historical and aerial photographs and data
obtained prior to and during the RI, it is believed that Honeywell may
have historically disposed of large quantities of Solvay wastes and other
chemical wastes directly into Onondaga Lake in the 1930s, 1940s, and
1950s."[53]

As would be expected, the highest concentrations of mercury were
found in the beds and waters of the creeks flowing into the lake, although some creeks had higher concentrations of other metals and or-
ganics than others. Also, the wetlands surrounding the areas were highly
contaminated with the same substances as the creeks and creek beds.
One example is the "dredge spoils area," a wetland into which Allied
threw contaminated material that was dredged from Nine Mile Creek
in the 1960s.[54]

Eventually of course, given gravitational pull, many of these wastes
ended up at the bottom of Onondaga Lake. Mercury contamination
there ran two meters deep and chromium contamination six meters.

Cadmium was, according to the report, "widespread" in the top two meters of lake bottom. Near the mouth of Nine Mile Creek, the cadmium reached eight meters deep. Other metals including lead, nickel, and zinc, were also widespread, reaching between two and seven meters into the lake bed. Organic compounds were found mostly to be along the shoreline, except for PAHs in the southern basin, like the residue from the Oil City facility.[55]

Because there were so many kinds of contaminants, and because the sources of the contaminates were so varied, tracking their entry into the lake and developing an understanding of how they interacted with it was a complex task, and the prospect of orchestrating a cleanup was daunting. Mercury was still flowing into the lake from some creeks; others were sources of organic compounds, PCBs, and heavy metals. Groundwater was a continuing source of organic contaminants of various kinds. Dredging the creek beds, while helpful, would not eliminate these other sources of pollution.

Ultimately the most significant problem was mercury. The report noted that "the lake sediments represent a huge source of mercury." As a result, even if the flow of mercury into the lake could be stopped, mercury pollution would be ongoing. While the mercury had settled into the lake bed, and sometimes perhaps deep into it, "it was not considered sequestered." Because of wave action and also lake stratification, that is, the shifting of water from lower to higher parts of the lake as the seasons change, materials at the lake bottom would constantly make their way into the lake's upper waters. A lake isn't stagnant. It is essentially alive and in constant circulation, picking up the sedimentary material from the lake bottom and transporting it. As the report noted, "internally derived loads of mercury impact all major regions (i.e., both the hypolimnion and epilimnion, i.e., the shallow and deepest) of the lake during the period of stratification."[56]

As a result of the stratification process, mercury levels can vary in different parts of the lake during different times of year. Moreover, methane bubbling up from the lake bottom pushes mercury into sediments and into the lake itself. Elemental mercury combines with carbon and hydrogen to form methylmercury, during a process known as "remineralization." As this occurs, mercury diffuses throughout the lake waters and permeates the entire ecological system. The grasses, trees, shrubs, fish, and smaller organisms become infused with mercury and other contaminants. In this way, the lake becomes a self-reinforcing system of toxic amalgamations.[57]

The lake's fish, for example, contained .02 to 4.7 milligrams per kilogram weight of mercury, depending upon their location (anything over .5 is considered potentially problematic). Their bodies also held dioxins and furans.[58] If that were not bad enough, "Honeywell-related contaminants and ionic wastes in Onondaga Lake have produced adverse ecological effects at all trophic levels examined." The macrophyte, or aquatic plant habitat, had been destroyed by all of the dumping, and its loss disrupted organisms up the chain, from invertebrates to vertebrates. All of the phytoplankton, zooplankton, benthic invertebrates, and fish had been contaminated with mercury, and often with the addition of other heavy metals and organic compounds. And since these kinds of contaminants persist for a very long time, they were "unlikely to decrease significantly in the absence of remediation."[59]

One of the RI's more disturbing findings involved the presence of 8,800 grams of mercury that was not accounted for by the calculations of inputs going into the lake from various sources. Far from insignificant, it constituted 70% more than the mercury "budget" that had been estimated in previous studies. More than likely, mercury was being released from the lake bed itself at higher rates than had been anticipated, perhaps from deep within it.[60]

Humans were deemed relatively safe from exposure, as long as they took precautions. The reported isolated twenty-eight pathways, apart from fish, which might potentially lead to human exposure. Twenty-one generated minimal health risks. The greatest was for older children who might be exposed to contaminated wetland sediments while fishing or wandering into the lake's sediments. Negative health impacts increased significantly for those subjected to multiple pathways of exposure, such as someone who walked into the lake's sediments to fish and then ate the fish that they caught.[61] The general rule of thumb seemed to be: don't get too near the lake. And, if you do decide to fish from it, you should probably be in a boat, or least wearing high waders.

When the report was published in 2002, people were fishing from the lake and had been for years. After all, it was a good place to catch fish. "The current fish population in Onondaga Lake," the report noted, was vastly improved over conditions in the 1950s, when over 90% of the fish in the lake were common carp (*Cyprimus carpio*)." The number had increased to over fifty species, among them bass, walleye, northern pike, and bluegill. It had become a destination of choice for bass fishing, with tournaments being organized, starting in 2001.[62] There were plenty of fish. You just shouldn't eat them.

Competing Clean-up Proposals and Visions of the Lake's Future

The Remedial Investigation (RI) was a prelude to Honeywell's long-awaited cleanup plan, which was released to the public in April 2005. The company promised that it was "committed to a process that includes sound science, regulatory review, and public participation—taken together, this will help eliminate pollution into the lake, enhance fish and wildlife resources, expand recreational uses and serve as a catalyst for future economic development."[63] The statement reflected the utilitarian view that had long been the impetus for lake cleanup. The lake was essentially an economic resource. Such a position leads, almost inevitably, to some form of cost-benefit calculation. How much should be invested in a restoration process, given the potential economic benefits that might result?

The proposal consolidated the fifteen contaminated sites described in the RI into eight Sediment Management Units (SMUs), each with its own remediation requirements. The designation carried with it a whole set of implications. The lake's beauty, its value, its history, and its geology are distilled into a series of discrete constituents, each to be subjected to differing levels of technical intervention and control. Such a schemata was a necessary prelude to the lake's potential restoration, but the necessity was determined by decades of industrial despoliation. The lake, its watershed, and the area's residents had traveled a very long and troubled path from the time when Peacemaker and Hiawatha negotiated the Great Peace in the vicinity of Onondaga's shorelines. Suppose that Hiawatha was to return and follow the path of the eight SMUs. What would he find?

Starting on the southwestern end of the lake, he would first encounter SMU 1, the area with the highest levels of concentration from the most diverse sources. Most of the in-lake waste deposits (ILWDs) were placed

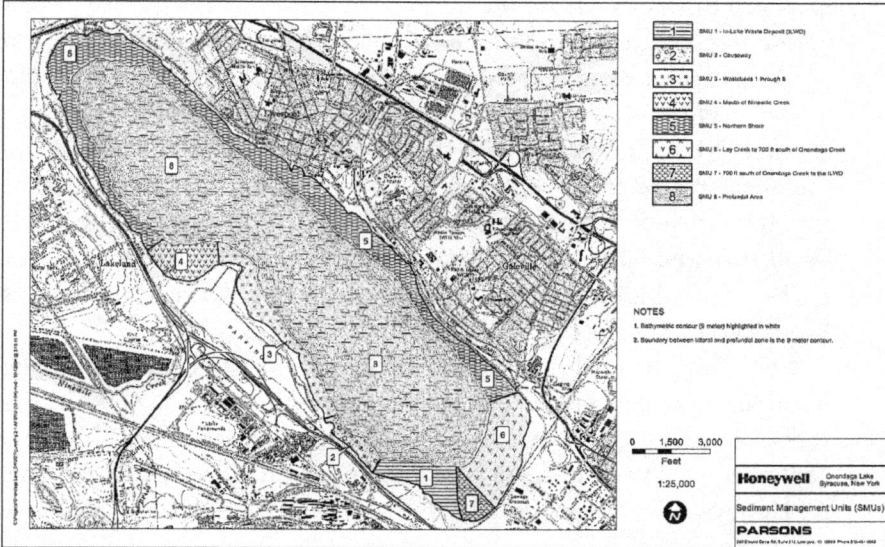

Figure 5.2. Eight Sediment Management Units (2002). Courtesy of the Honeywell Onondaga Lake Cleanup website.

here. A large delta was created that "included a combination of cooling water, sanitary wastes, Solvay waste, mercury wastes, and organic chemical wastes, which settled out at a higher elevation than the surrounding areas of the lake bottom." Given the volume of material and its soft character, the delta had a history of landslides, indicating an instability that might prove to be a problem for capping.[64]

Hiking northward, Hiawatha would encounter SMU 2, which stretches 3,000 feet along the shoreline and several hundred feet into the lake itself. The area was used by Honeywell for loading and unloading materials into its various facilities. When Interstate 690 was constructed in the early 1970s, waste materials from the construction process were deposited here. Primary contaminants include chlorinated hydrocarbons, non-aqueous phase liquids (i.e., petroleum and its byproducts), mercury, and other heavy metals.[65]

Next, trekking northward, Hiawatha would have to hike up a hill that had not existed during his lifetime. Now referred to as SMU 3, the hill is dominated by a large Solvay wastebed, composed primarily of calcium carbonate, with various toxics intermixed. These materials had been disinterred from the limestone deposits generated over a vast period of geological time, which had been removed and pulverized to produce carbon dioxide for soda ash production, and then piled onto a giant slag heap. Hiawatha, still walking northward, would next encounter SMU 4, associated with the outflow from Nine Mile Creek and the delta created from wastes dumped into it. Sources of contamination here included the LCP Bridge Street Site, the West Flume, Geddes Brook, and Nine Mile Creek. The West Flume was a repository for LCP's chlor-alkali production wastes, which then flowed into Geddes Brooks and Nine Mile Creek.[66]

Continuing around the lakeshore, Hiawatha would be walking along what was now designated as SMU 5, which runs from SMU 4 all the way around the lake until it abuts SMU 6 at its southern end. This area contained the lowest levels of contamination, but it was hardly pristine. As Honeywell's analysis noted, "The sediments through the SMU are dominated by calcium carbonate and oncolites. Run-off from industrial areas, suburban developments, and roadways continue to contaminate this area with a variety of toxics."[67]

Hiawatha's journey would terminate back at the southern end of the lake, at SMUs 6 and 7. SMU 6 was the site of the Oil City storage facility, a GM dredging site, the Town of Salina landfill, and wastes from the Erie Boulevard Manufactured Gas Plant, Roth Steel, and American Bag and Metal. Although Allied was not in the immediate vicinity, evidence of its wastes can also be found here. The sites contain an assortment of organic chemicals, mercury, and arsenic.[68]

SMU 7 is squeezed between Harbor Brook and Onondaga Creek. The dominant contaminants here are ILWDs (in lake waste deposits), probably the result of residue from Oil City. The prevalence of hydrocarbons became apparent when a sheen appeared on the lake's surface after borings were taken from its sediment. As scientists expected, toxic organic compounds were present, along with heavy metals.[69]

SMU 8 could be observed, if at all, only from a canoe. It included the prefundal area (the deeper part of the lake). While few, if any, toxics were dumped directly into this zone, "The ultimate fate of most of the sediment entering Onondaga Lake is burial in the prefundal sediment." That is, contaminated sediments were slowly and inevitably pulled down by water and gravitational forces into the deepest parts of the lake. The prefundal areas are anoxic (devoid of oxygen) in the summer months, a phenomenon common to stratified lakes, but also exacerbated by the toxic load that burdened Onondaga.[70]

Clean-up Options and Objections

Honeywell put forward seven potential plans for the cleanup. The smallest intervention, Alternative 1, involved no action. The greatest intervention, Alternative 7, involved isolation capping of 1,826 acres of the lake bed, with the entire lake bed having a thin cap.[71] Alternatives 2–7 involved some aspects of each of these actions. The differences among these alternatives tended to revolve around the scope of dredging and capping, and the determination of what would be dredged was related to measurable levels of contamination in the sediment. Honeywell also calculated the costs of each alternative and contended that alternatives 5–7 were not worth the increased costs, given the reduced levels of toxics that would result. Alternative 7, for example, was estimated to cost $2 billion and take seventeen years to complete.

Honeywell first proposed a version of Alternative 2, which would involve minimal dredging and capping. The state rejected it. DEC officials contended that it did not adequately address the issue of mercury in sediment at the lake bottom.[72] The state seemed to opt for Alternatives 6 or 7, which involved dredging the entire lake and putting a cap of sand and gravel over the areas which were most heavily contaminated.[73] The DEC then released a more detailed version of what it had in mind. It involved removing the entire lake bottom to a depth of 30 feet, and then putting a cap over the 425 acres of the most contaminated area. The cost: $2.3 billion.[74] *The Post-Standard* editorialized in favor of the DEC's proposal: "Community leaders and the DEC must not be swayed by Honeywell's schmooze sessions. It is absolutely critical that the cleanup is done right, based on scientific certainty."[75]

The DEC's proposal was not well received by Honeywell. The company began a lobbying campaign that targeted not only state officials, but also business leaders and environmentalists. The company's messaging included veiled threats of resistance should the state push too hard. Gary Brown, from the engineering company that was working with Honeywell on its proposed plan, discussed the company's persuasive strategy. "The message was, 'Don't let this opportunity slip away . . . The idea is not to debate the science so much as to make people aware of the issues.'"[76] Before long, their efforts seemed to pay off. Soon state officials were publicly supporting something more modest. Their plan involved dredging only the contaminated part of the lake bottom and placing the cap over it. The cost, at $449 million, was twice what Honeywell had proposed, but still about five times less than the DEC's original plan. The state was now supporting what was essentially Alternative 4.

The proposal involved dredging the most contaminated parts of the lake, with the dredged material being transferred to an area three miles uphill from the site. Wastebed 13, in the town of Camillus, was suggested as the optimal location, whose residents would have to deal with the consequences. Isolation capping in dredged areas would consist of an "armored layer," of sand and gravel, over which would be laid a six-inch "habitat/bioturbation" layer that would support various forms of aquatic life. Thin layer capping was proposed for parts of the lake's profundal zone. The intention was to create a layer on the lake bottom that would block contaminated sediments from making their way into the lake's waters. Aeration would be introduced to further reduce the level of methylmercury leakage. This would require ongoing intervention for an indeterminate period of time. Monitoring of the sites would also continue indefinitely into the future.[77]

The state's compromise proposal was now coming under fire from cleanup proponents. Scientific critics charged that this plan would not actually lower mercury levels in the lake, and the Onondaga Nation released a statement that the plan was insufficient to restore what they considered to be the lake's sacred waters.[78] Representatives of the Onondaga Nation had a different vision of the lake's significance that was incompatible with the utilitarian calculations being made by the state. Their response would mark the beginning of a long-running dispute over the adequacy of lake cleanup efforts. Eventually the Onondaga Nation would split with some area environmentalists, their former allies, on the issue of how extensive the cleanup process should be and how it should go forward.

In June, Honeywell sponsored a poll conducted by Charlton Research Company, the putative purpose of which was to gauge the attitudes of area residents on its proposed cleanup plan. But the questionnaire was apparently a push poll, which did not allow a negative response (a common strategy for political campaigns). Roseanne Corrigan, a resident of Camillus, said, "What struck me as funny was that they didn't give you an option to answer that you disagreed. The options were to answer if you 'strongly agree' or 'somewhat agree.'" When the poll worker told her that those were the only options, she asked that her name and responses by pulled from the poll. A Honeywell spokesperson, when asked about the poll, said that "Every person who participated was given a whole spectrum of potential responses." When Corrigan asked the poll taker who was sponsoring the poll, she was refused an answer, but she became convinced it was Honeywell. One question was whether she wanted to see a "clean lake." Her response: "Of course, I want to see the idiot who wants to see a dirty lake."[79]

Recognition of the complexities and interconnections of long-term contamination continued to grow even as plans for a cleanup were moving forward. In summer 2004, researchers discovered brine fields moving toward the lake in groundwater. Brine had been a key component in the Solvay soda ash process, and the fact that Syracuse had large salt beds was a key reason that the Solvay plant was located in the area. Huge amounts of salt had been removed from the Tully Valley during Solvay's period of operation. The salt was mixed with water to create brine and then pumped to the facility, but large amounts of residual brine remained and were now slowly moving into the groundwater and then to the lake. The state was tracking the brine, and while it would likely not leach into the lake itself, given its heavier density and position below the lake, there was concern that it could react to chemicals that pervaded the area and exacerbate existing toxics problems.[80]

As different alternatives to remediation were considered, it became clear that even a modest dredging operation posed a disposal problem. What would be done with the thousands of tons of muck that was going to be pulled out of the lake? The state was proposing that it be piped to an old industrial landfill in the town of Camillus. Residents and public officials were understandably skeptical about such a project. Odors, and the toxics that they represented, were of special concern, given that a new housing development was slated for construction nearby. When the old landfill had been in use, the area was sparsely populated, but now it was an expanding suburban community. Camillus officials wondered why the removed detritus couldn't be deposited close to the lake itself.[81] The Camillus disposal operation would eventually become the subject of active grassroots opposition and legal challenges.

The dredging and capping operation was only part of an incredibly complex plan to deal with multiple contamination issues. Honeywell had also put forward a $20 million plan to clean contaminated groundwater that was still seeping into the lake. It involved a constructing a filtering facility, effectively putting an end to continued toxic discharges from the old Solvay facility.[82]

In spite of the complexities involved and the potential for setbacks, a sense of optimism was beginning to emerge at least in some quarters, that if Onondaga Lake's waters could not be completely restored to a pristine state, they could at least be significantly improved. In August 2005, *The Post Standard* editorialized that "scientists were downright giddy over the latest water-quality test results from the lake." The article was referring specifically to another major piece of the cleanup operation—the expansion and upgrade of the Metropolitan Sewage Treatment Plant. The city and county had taken the first steps in what would be a fifteen-year operation, one that would cost nearly $500 million. This would be, the editors noted, "the first step toward transforming Onondaga Lake into the regional economic and recreational hub it rightfully should be."[83] Nevertheless, the utilitarian rationale for cleanup efforts, while appealing to local businesses, public officials, and many members of the public, was anathema to members of the Onondaga Nation.

In September, Honeywell announced a significant step forward in reducing mercury contamination. They had cut off what they said was the "largest source of mercury leaking into Onondaga Lake." The company had removed seven tons of mercury from the property once owned by Allied Chemical and then subsequently to LCP Chemicals. LCP had used the facility to make chlorine, which, like the Solvay process, needed large quantities of brine. The state intervened to shut the facility down in 1988 due to repeated chlorine gas leaks. Four years later LCP declared bankruptcy. It took Honeywell two years to clean up the thirty-acre site, using a relatively new "soil washing" technique. The process resulted in 19,000 gallons of mercury-laced liquids and 280,000 pounds of toxic sludge, which had to be disposed of someplace. The company also built a 100-foot underground wall to keep any residual mercury from leaching into the lake. Company official John McAuliffe declared it a "major milestone."[84]

At the fifth annual Lake Progress Meeting, state officials were able to tout several markers of improvement in the lake. Sport fishing was returning to the lake, ammonia levels were dropping, and Honeywell had finally cleaned up the LCP site. But local residents who attended the meeting still had plenty of complaints. Weeds and algae were continuing problems. And raw sewage was still being dumped into Onondaga Creek.[85]

In October, SUNY ESF (College of Environmental Science and Forestry) faculty held a public forum and tour of the lake at which they also praised improvements to its water quality. At the forum, Honeywell explained its plan to fulfill the state's mandate for a $451 million cleanup. Dick Elander, director of Onondaga County Water Protection, noted that with half of its $460 million share of the cleanup, twenty-five projects, mostly having to do with improving sewage treatment and reducing discharges, had been completed by the county. Representative James Walsh, who had been instrumental in securing federal dollars for the county's share of the cleanup, said, "It's just so exciting to see us taking the lake back."[86]

Honeywell was also moving forward with the underground barrier designed to keep contaminants from the LCP property from reaching the lake. Construction was visible from Interstate 690, a reminder to the area's residents that the lake cleanup operation was underway. At a cost of $20 million, the system, designed to divert groundwater away from the lake into a water treatment plant via a system of pipes, was relatively inexpensive in relation to the project's entire cost of $451 million.[87] By the end of November, Honeywell had completed the 550-foot pipe that would drain 200,000 gallons of contaminated groundwater designated to be moved to the proposed treatment plant. The company extolled the project as evidence of their commitment to cleaning up the lake's pollution.[88]

Members of the Onondaga Nation were still skeptical of the state's proposed cleanup plan and Honeywell's actions. The Nation objected to the partial dredging of the lake bottom and expressed doubts about removal of mercury from the old LCP site. Joseph Heath, a lawyer representing the Onondagas, was highly critical of the company's mercury remediation efforts. According to Heath, only mercury on the surface of the ground near the facility was being removed, leaving contaminated soil beneath it, eighteen to fifty feet deep. The state disputed Health's claims, with Ken Lynch, DEC regional director, suggesting that the Onondaga's estimates of residual mercury were "based on erroneous assumptions."[89]

In May 2005, the Onondaga Nation announced that it had hired a high-profile consulting firm, Stratus Consulting, to take a look at Honeywell's proposed cleanup effort. Stratus Consulting had worked on the Hudson River PCB contamination case and for the Port of Houston, which won $100 million from nearby chemical companies. The Onondagas were calling for a complete dredging rather than a cap, a project that Honeywell suggested would cost $2 billion.[90]

In July, however, a federal judge refused to allow the Onondagas to delay approval of the DEC's final decision. As a result, the Stratus analysis would not be completed in the time to be included in the state's recommendations. Judge Fredrick J. Scullin would not even allow the Onondagas' attorney to put forward arguments in his court. Attorney Joseph Heath was incensed. "This is not a proper remedy and is not compliant with the law," he argued. "There are rights here that are being trampled and ignored. And it's not just about the [Onondaga] nation. All the people in Central New York are being shortchanged in a big way." Heath was skeptical that the state's plan would actually clean up the lake. "They would leave so many toxics in the bottom of the lake that it would still be a Superfund site."[91]

The Onondagas wanted to bring the public's attention to Honeywell's reluctance to commit to the lake cleanup, so they put forward a resolution to Honeywell shareholders that would have required the company to put up warning signs, alerting the public to the dangers of eating fish caught in the lake. While the dangers of ingesting toxins with any fish from the lake were real, the resolution itself was rhetorical, designed to publicly shame the company and encourage them to expand and speed up their cleanup efforts.[92]

At about the same time, the state suggested a $451 million plan, taking what amounted to the middle ground and signaling a desire to put an end to a sixteen-year legal battle. But a Honeywell spokesperson was noncommittal. Honeywell Director of Communication Victoria Streit-feld stated that "The DEC's selection of a final plan and its release of the record of decision represent continued progress on the cleanup of Onondaga Lake." The state had made concessions to Honeywell, but the company would have to sign off on the plan before it could be put into effect.[93]

As 2005 ended, the two parties agreed to a major element of the re-mediation plan involving mercury remediation. Under this plan, 2.64 million cubic yards of dredged lake bottom material would be removed from the lake bottom and a cap placed on 579 acres of lake bottom. This would constitute the most expensive environmental cleanup in the his-tory of New York State. Once Honeywell signed off, it would put to rest a lawsuit that was initiated by the state in 1989.[94] Final details of all ele-ments of the lake's remediation had yet to be resolved, but it appeared to be only a matter of time before they would be addressed.

While controversies, conflicts, and negotiations continued in con-nection with the industrial contamination sites in the area, legal cases were moving forward on the Metro sewage plant treatment front, spe-cifically related to the Midland Avenue Regional Treatment facility. The Partnership for Onondaga Creek (POC) filed a complaint with the U.S. Environmental Protection Agency (EPA) Office of Civil Rights (OCR) contending that Onondaga County and the State of New York Department of Environmental Conservation (DEC) violated the Civil Rights Act of 1964 by approving the Midland plant. The POC alleged in their complaint that their exclusion from "meaningful participa-tion" in the siting process, involved a "continuing pattern of discrimi-natory conduct." Moreover, they contended that the construction and operation of the Midland facility would "have adverse disparate impacts upon the predominantly African-American residents" of the community.[95]

In a letter to Alma L. Lowry, Director of Public Interest Law Firm, who had represented the POC, the EPA rejected both complaints. The first was dismissed on procedural grounds, in that it was not filed within the 180-day window between when the discriminatory action had oc-curred and the deadline for filing a case. The adverse impact issue was taken up by the OCR but given short shrift. The OCR accepted the 1999 EPA review of the plan as carefully executed and noted that the size and scope of the original facility had been reduced. Given this, they found no evidence of an adverse impact on the predominantly African American community.[96] At this point, it is worth noting that residents on two of the streets near the proposed facility, Oxford and Blaine, had already been relocated, so impacts on the community were felt in a very tangible way, a fact not recognized by the OCR.[97]

In 2006, after spending several years moving through state and federal courts, a federal district court found in favor of Onondaga County's authority to condemn the Midland Avenue site and begin construction of the RTF. The case was decided on jurisdictional and process grounds involving the New York State law and the Onondaga County Administrative Code (OCAC). In short, the OCAC "does not require the County to obtain Common Council approval before making any final determination regarding the establishment or extension of any sewage project with or serving Syracuse."[98] With this, the County was prepared to go forward with construction of the project.

In June 2006, a new source of contamination for the lake was discovered leaking from the Solvay wastebeds. At issue were the tons and tons of calcium chloride that were disposed of by the old plant after it stopped dumping it into the lake. The beds covered 300 acres on the western side. While the calcium chloride is not in itself toxic, it was laced with a stew of other contaminants, including benzene, toluene, xylene, naphthalene, and phenols. The findings regarding this new problem, outlined in a report by the DEC, complicated the cleanup plans.[99]

That fall, the State of New York Environmental Science and Forestry School proposed planting shrub willows on the Solvay wastebeds. Proponents of the plan, based on research funded by Honeywell, suggested that the fast-growing willows would provide a cover over the 662-acre area and prevent the calcium chloride from leeching into the lake, thus allowing the return of plant life to the nearby shoreline. Trees would be planted directly into the wastebeds and fertilized with sewage sludge from the nearby Metropolitan Sewage Treatment facility. The trees would be cut down every three years and used as fuel in a proposed power plant. If the project were successful, it would save Honeywell the millions of dollars that it would cost to construct a cap to cover the area, giving the company an excellent return on their $300,000 grant to the SUNY ESF researchers.[100]

In the midst of all the controversies related to various restoration efforts, a series of land claims cases found their way through New York State courts, which had the potential to upend agreements being negotiated between Honeywell and the State of New York. Haudenosaunee claims are grounded in illegal land expropriations that began with the infamous Sullivan campaigns of 1779 during the American Revolutionary War, and which continued via questionable treaty agreements that transpired afterward. A whole body of case law was established in the twentieth century, which in general worked to find legal justifications to deny Haudenosaunee access to courts, and when access was granted, to minimize the legitimacy of any potential claims. In spite of this, the determination of Indigenous communities in New York to find legal redress for their arguably legitimate grievances seldom waivered. The Cayuga Nation, for example, waged legal battle to have its claims legitimized, beginning in the 1980s into the twenty-first century. In 2000, a federal district judge allowed a jury to be selected and a trial to go forward in order to consider a Cayuga land claims case against George Pataki, Governor of the State of New York. The jury found in favor of the Cayuga Nation and awarded $247 million in restitution. The award for the first time seemed to open a window to a broader set of claims by the Haudenosaunee across New York State. The decision also sent a shock wave to white landowners, especially in southern Central New York where Cayuga lands were located, which was underscored when the verdict was upheld by a district court judge.[101] In response to the decision, in 2004 Governor George Pataki offered the Cayuga a casino license in the Catskills, which they accepted.

The Cayuga jury award encouraged other members of the Confederation to bring land claims, and among these was the Onondaga Nation. The Onondagas made national news in 2005 with their decision to sue in federal court to claim the land underneath the City of Syracuse and surrounding areas, which included several thousand square miles. The objective of the claim, however, was not to obtain the land, but to secure environmental protection. Central to the Onondaga's claim was the restoration of Onondaga Lake. Sidney Hill, a leader of the Onondaga, made their priorities clear: "Our concern is for the water, the land, the air. They are not well. It is the duty of the nation's leaders to work for a healing of this land, to protect it, and to pass it on to future generations." Hill characterized the Cayuga–Pataki settlement as a "divide and conquer" strategy, with the governor attempting to essentially buy off some members of the confederation at the expense of others.[102]

The case had potentially far-reaching implications. Had it been successful, it might have provided a legal mechanism for Indigenous people in New York and possibly other states, to use land claims as a means to require a broad range of environmental protection and restoration efforts. More specifically, it would have given the Onondaga leverage with the New York Department of Environmental Conservation to intensify their restoration efforts. As the Onondaga claim proceeded, however, a federal appeals court nullified the Cayuga claim and settlement on the grounds of "laches." Laches is to civil law what the statute of limitations is in criminal cases. The appeals court determined that the Cayuga had waited too long to file their claim, and as result their right to do so had expired.[103]

Onondaga Chief Sidney Hill responded in an op-ed in the *Syracuse Post-Standard* that Native Americans did not have access to state courts in the eighteenth and nineteenth centuries; that they were "victims of constant political, social, economic exclusion and oppression"; and that the U.S. government had promised to protect their land claims. In the 1920s, the Onondaga made various efforts to exercise claims, but the courts ruled that federal courts didn't have jurisdiction. Not until 1974 were Native Americans given access to federal courts, and only in 1985 did the courts rule that some claims might be valid.[104] Given that Native people were denied access to American courts to file land claims through most of the twentieth century—not for lack of trying—the decision rests on questionable grounds, and the laches argument seems disingenuous. The precedent set by the appeals court in the Cayuga case was used as grounds to nullify the Onondaga land claim as well.[105] As a result, the small window that was opened up in New York state for Indigenous people to restore their illegally expropriated lands, to use their rights to them in the interest of ecological restoration, was closed. The U.S. Supreme Court refused to hear either the Cayuga or Onondaga cases, thus affirming the appeals courts' rulings.

In September, famed primatologist Jane Goodall threw her support behind an extensive lake cleanup. She stood in solidarity on the shores of the lake with members of the Onondaga Nation to express her concern for its destruction and her hope for its restoration. In a ceremony held on the lakeshore, a group of Onondaga children poured an urn of clean water into the lake as a symbol of its return to health. Tom Porter, a Mohawk spiritual leader, said, "We are in a way like the lake. We've been colonized. We've been polluted. We've been hurt. We can change it, we can fix it, but we have to know what happened." And Onondaga Faithkeeper Oren Lyons stated, "If we could clean this lake, it would represent one of the greatest pictures of hope we can give the world."[106]

At the same time, complaints were being leveled against both the state DEC and Honeywell over the secrecy of negotiations to finalize the approved plan for the full lake cleanup. Joseph Heath, and attorney representing the Onondaga Nation, was particularly dismayed, stating, "We have a major decision for our community—this lake and its health and whether or not it gets cleaned up, is one of our biggest issues—and the meetings are absolutely secretive. This is not the way I imagined a democracy should work." Ken Lynch, the DEC regional director, on the other hand, responded that, while current negotiations were held behind closed doors, all agreements would be subject to public hearing, and there would be opportunities to respond. "The clean-up," he said, "has and will continue to be, a public project."[107] Clan Mother Wendy Gonyea told me that officials from the EPA visited Onondaga territory to give a presentation of the R.O.D. They never asked for input or advice. The decision was a *fait accompli*, a formality. Government officials came to check the bureaucratic box.[108]

As progress was being made toward attempting to correct the century-long despoilation of Onondaga Lake, optimism that the cleanup would be expeditious or uncontested would soon be eroded by the complex technical, political, and legal challenges posed. A key feature of the cleanup was the authority given to the New York State DEC to negotiate a settlement with Honeywell. This resulted from the DEC lawsuit, in which the company agreed to settle with the State of New York (and the USEPA). The lawsuit, in turn, stemmed from the legal authority given to the DEC by the state to address such cases. So there was legal basis for this outcome. But the result was a fairly centralized decision-making process. A success for the Onondaga Nation in its land claims case would have opened up the process and given its representatives substantial input into decision-making. Its failure reinforced the political marginalization with which they had been treated for decades, if not centuries, and it reinforced a sense of exclusion.

6

A Sacred Lake, a Recreational Resource

Competing Visions

In the fall of 2006, Honeywell began sinking steel pilings into the Onondaga shoreline, the first step in constructing a barrier to keep mercury and other contaminants from leaching into the lake from the old Solvay site. However, the discovery of a sewer pipe and a series of wires forced an alteration in the plans. Now sections of the steel barrier would be placed inside the lake itself, shrinking the 3000-acre lake by two acres. Members of the Onondaga Nation objected. Jeanne Shenandoah stated, "Putting the wall into the lake like that, I don't think it's going to make things any better. It would affect fish and wildlife." But DEC director Ken Lynch was, as usual, more upbeat, saying, "We have the opportunity to redesign a shoreline." DEC officials suggested that the lake could be expanded in other areas, although how exactly that might be accomplished was not specified.[1]

In October, Philip Arnold suggested in a column in *The Post-Standard* that the Onondaga people and the state had fundamentally different views on the value of the lake. The state saw the lake primarily as a *"recreational"* resource, whereas members of the Onondaga Nation, and the Haudenosaunee viewed it as sacred, with deep-rooted cultural significance. These differing views led to different conclusions as to what needed to be accomplished.[2] The Syracuse Chargers Rowing Club, for example, endorsed the DEC and Honeywell's plan to clean up the lake on the grounds that members wished "to see sporting and recreation activity continue to grow and thrive on its waters." Rowing, of course, as the letter pointed out, does nothing to undermine a lake's water quality, but club members, like other non-Indigenous members of the community, seemed generally to have a more utilitarian view of the lake's uses than members of the Onondaga Nation.[3]

The Atlantic States Legal Foundation, one of the original plaintiffs in the case against Honeywell, also tentatively endorsed Honeywell's proposed plan in December. Foundation president Samuel Sage stated in a letter to the *Post-Standard*, "For a variety of reasons, we at the Atlantic States Legal Foundation feel the plan is not perfect. But we are convinced that the plan is a good start and that knowledge gained during the five year-design stage will allow us to hone in on what exactly needs to be done."[4]

Nevertheless, 2007 began on a less optimistic note when the DEC and Honeywell were unable to agree on standards for cleaning up Geddes Brook and Nine Mile Creek. Honeywell had agreed to develop a timetable for a cleanup of the lake's tributaries, but U.S. District Judge Frederick J. Sculin granted a twenty-two-month extension of the agreed-upon January 8 date, because the parties could not decide on the most effective plan. This again pointed to the intertwined nature of the various contamination problems and undermined the larger cleanup effort. It wouldn't make sense to dredge the lake bottom until contamination from the creeks was effectively addressed. DEC director Lynch was optimistic as always, suggesting that the extra time would allow for testing that could result in the state and Honeywell finding common ground. Moreover, since dredging operations were not slated to begin for five years, the delay would not negatively impact the overall lake cleanup. Onondaga Nation legal staff were less sanguine. Thayne Joyal expressed concern that site-specific testing of the waters would lead to more lenient standards. The Onondagas wanted a more complete package of tests.[5] But in a letter to the *Post-Standard*, Honeywell official John McAuliffe contended that Honeywell was ahead of schedule in the cleanup process. He noted that the groundwater treatment plant in Geddes was completed a year ahead of schedule. The predesign for the lake cleanup was underway, and the company was close to submitting its Remedial Design Work Plan to outline how the entire process would be implemented.[6]

Although things were moving forward in fits and starts and not without some controversy, some progress was undeniably being made. Given that, the question of who deserved credit for it started to seem pertinent. Nick Pirro, Onondaga County Executive, was receiving a fair amount of the credit. SU Professor Charles Driscoll stated that Pirro "was forced to do it, but there are some very innovative techniques being used in the wastewater treatment plant. The bottom line is that there have been remarkable improvements." It was during Pirro's term in 1997 that the County Legislature approved the original $380 million plan to improve the Metropolitan sewage system. Pirro was Executive in 2006, when the $461 million agreement was reached between the state and Honeywell.[7] But Samuel Sage, who sued the county to address the sewage treatment plant issue, was less generous. "He was extremely negative toward spending any money on the lake . . . It was always antagonistic."[8] Pirro stepped down in 2007 after having served as county executive since 1988.

SUNY ESF was fast becoming a world leader in studying and promoting the use of willow shrubs as an energy source and for environmental remediation. They declared the experiment with willow shrub growth on the Solvay wastebeds, which had begun in 2006, to be a success. Not only were the shrubs providing ground cover, but they were sucking up contaminated water that might otherwise flow into the lake. Researcher Lawrence Abrahamson noted the utility of the shrubs as a renewable resource: "We look at a ton of willow like a barrel of oil . . . You can do a lot of different things with it."[9] The project started with sewage sludge as the fertilizing agent, but eventually, waste from an Anheuser-Busch Brewery facility in Lysander was added to the mix to make artificial soil that proved ideal for growing the shrubs.[10]

In July, the county proposed extending the Onondaga Lake hiking path for two miles along the shoreline. "It's the most amazing portion of the trail," according to the Parks Commissioner, Robert Geraci touted the extension. He suggested that this part of the lake seemed completely isolated from the rest of the city and if you were dropped into it "You wouldn't know where you were."[11]

In October, Honeywell officials held a ceremonial tree planting at the old LCP site to mark the completion of its transition to a restored wetland.[12] Eight tons of mercury had been removed from the site before the tree planting began. 12,000 plants were added, drawn from twenty native species that included everything from pussy willows to the carnivorous water plant known as bladderwort.[13]

The headline in the *Syracuse Post-Standard* was ecstatic: "Lake Lives Again!" The year 2007, it declared, involved the lake's "biggest improvement in 20 years," adding that it "reached its highest oxygen levels in at least 40." This improvement was primarily because of the completion of the upgrade to the Metropolitan Sewage Treatment Plant. The dead zone at the lake's bottom, an area completely devoid of oxygen during a period of time every year, had been reduced from five months in the 1970s to three when the article was written. The smell of the lake had improved because ammonia and phosphorus levels had declined. Mercury levels were also slowly decreasing, a trend that was evident in the lake's zooplankton. But the improvements had not yet made their way up the food chain into fish. So, while improvements in the lake seemed to be progressing, using the lake for edible fishing or swimming was not advisable. Steven W. Effler, a researcher at the Upstate Freshwater Institute, maintained the lake was still among the most polluted in the United States.[14]

Efforts weren't just about the lake itself, but also about continuing to restore the land area adjacent to the lake. Tony Eallonardo, a doctoral student at SUNY ESF, was engaged in a project to restore salt marsh plants that once thrived in the briny springs that ringed the circumference of the lake. The plants were gone, and only one salt spring could still be identified, but Eallonardo believed that some of the indigenous plants could be grown in the Solvay wastebeds. He and his wife Amber had established a test plot of 20,000 plants, and they seemed to be thriving.[15]

As the year ended, Honeywell began installing twenty-five-foot steel sheets along the western shore of the lake, designed to keep mercury from leeching from the LCP facility. It would be followed by the placement of pipes that would carry the groundwater to a processing facility for removing toxics, after which it would be fed to the lake.[16]

In September 2008, reports of improvement continued to appear. The Metropolitan Sewage System upgrade was having a positive impact. Phosphorus was declining at a faster than expected rate, and the result was a decrease in algae and an increase in oxygen levels. The distinctive and very unpleasant "Onondaga Lake smell" was decreasing in intensity. Nitrate levels in the lake were also decreasing. Yet, in spite of these improvements, the lake still had serious contamination problems. In comparison, for example, Onondaga Lake had seventeen micrograms of phosphorous per liter, whereas Skaneateles Lake, from which Syracuse pulled its drinking water, was at three or four micrograms per liter.

Syracuse University Environmental Systems Professor Charles Driscoll seemed quite pleased with the progress that had been made. "If you asked me two or three years ago," he stated, "about the extent the lake would recover, I would never have predicted it. It's nothing short of spectacular. It's really outstanding."[17] And an editorial in *The Post Standard* was equally effusive. Not only was the lake cleaner, it suggested, but, "Far from bankrupting the county, this cleanup has been accomplished with more than $160 million in federal aid," to which it gave credit to Republican Congressman James Walsh.[18]

Still, dredging operations to remove mercury from the lake bottom had not yet begun. Honeywell's dredging plan targeted hundreds of thousands of yards from the lake bed. Sand and gravel caps would then be placed over the dredged areas. Operations were expected to begin in 2012. Honeywell was also in the process of completing a 1.5-mile underground barrier wall to keep more toxics from leaching into the lake from contaminated sites on the shoreline.[19]

A study done by the Smithsonian Environmental Research Center determined that mercury being currently released into the lake was more harmful than legacy mercury that had settled onto the lake bottom. The finding was based on a test study conducted in a lake in Northwestern Ontario, in which mercury was released into the water with fish already in it, which were then tested for mercury levels. Over time, chemical transformations resulted in the mercury becoming less available for absorption, and mercury levels in the fish declined. This was potentially good news for Onondaga, because it suggested that if mercury leaking into the lake from onshore sources could be eliminated, then mercury levels in fish could be improved relatively rapidly, even before the dredging operations were completed.[20]

Work continued on planting willow shrubs in the 700-acre Solvay wastebeds. ESF scientists and Honeywell officials took local community leaders on a walk through the ten-acre plot of newly planted trees. In phase two of the project, scientists were trying to find to proper mix of fertilizer with the calcium chloride residue that constituted the bulk of the material in the beds. With the latest planting of 60,000 additional trees, the project had reached a total of 100,000. Project organizers hoped eventually to create a massive living cap over the entire area of beds. Given the willow shrubs' tendency to absorb water, researchers hope that they would help prevent further contamination of the lake from the beds, while also providing a source of biomass for generating electricity.[21]

Some researchers were now also experimenting with planting native species on the beds. On one 5.2-acre parcel, fifty species of plants had been introduced, some of which were indigenous saltwater species that had been eliminated from the local area over the decades of toxic dumping. American elms were also being reintroduced at one plot, and another included white swamp oaks. In a somewhat ironic turn of events, the Solvay wastebeds become a kind of experimental station for observing conditions under which native plants might thrive. Don Leopold, a professor at SUNY ESF, said that he predicted the "area will be one of the most interesting sites in Central New York" for ecological observation and experimentation.[22]

Contamination of the lake and surrounding ground had not been the only problem generated by the Solvay Process. Sinkholes were becoming an increasing problem in Tully. While sinkholes were a natural phenomenon in the area, they were being exacerbated by salt-mining activities conducted by Allied Chemical throughout much of the twentieth century. The Solvay Process was located in the Syracuse area partly because of the presence of salt, but the nearby salt springs had been depleted by decades of extracting it. Needing the sodium from salt, the company turned to the Tully deposits, located about twenty miles south of Solvay. The deposits there were 1,000 feet below the surface, so engineers set up a system of injecting water, turning the salt into brine, which was then pumped to the process facilities. As a result, large cavities now existed in what had formerly been bedrock. The ground above soon started to sink, five feet in some places and as much as seventy in others. The largest hole had been nicknamed "The Big Sink," a large depression that had filled in with water. The bedrock, at least according to Honeywell, had stabilized, but fissures had recently appeared in Tully. Evan Van Hook, a spokesman for Honeywell, stated, "I think probably some of the fissures are related to subsidence from the mining . . . But nobody really knows how this happens."

An additional problem involved the "mudboils" that infected the area. The boils were created by pressure that accumulated in the collapsed bedrock, pushing water, clay, silt, and sand to the surface in small volcano-like eruptions. Ironically, "The Big Sink" was seen as providing a possible solution to the problem, because if it were to be drained, the water that provided the power for the sinkholes might dissipate to some degree. Ed Michalenko, president of the Onondaga Environmental Institute in Syracuse, had created a pilot project to use pumps, powered by wind and solar, to drain the water. Previous projects involved diverting water from the mudboil area, drilling wells to reduce underground pressure, and constructing a dam to divert silt. Combined, they had some impact, at a cost of $165,000 a year.[23]

By the end of 2008, the proposal for cleaning up Nine Mile Creek had become a flash point. The DEC and Honeywell had agreed to dredge the creek and place a cap similar to the one that was being proposed for the lake bottom, at a cost of $20.2 million. The state was proposing to move the contaminated material to the old LCP Chemicals site, where it would be contained in the same way that the LCP materials were. Some critics of the plan, however, wanted the debris trucked to a disposal site near Rochester, requiring an estimated 17,000 truck trips to remove it. State legislator David Scott favored an alternative plan, which would cost $10 million more and would involve deeper dredging, removing the necessity of a cap. County legislator James Corbett supported the plan, stating, "I couldn't in good conscience say we should take 17,000 loads up the Thruway when we have state-of-art facility less than three miles away."[24]

County legislator David Stott had serious doubts about the plan to dredge and cap in Nine Mile Creek. His reservations were partly based on the experience with capping that occurred in connection with a landfill in nearby Salina. According to Stott, in 1982 the Salina landfill was capped at a cost of $20 million, but, now twenty-six years later, the DEC had determined that the cap was insufficient to keep contaminants contained and was requiring the town to recap it at an additional cost of $26 million. Given that one of the main issues was mercury contamination, potential problems posed by capping at Nine Mile Creek and in Onondaga Lake itself seemed particularly acute.[25]

In February 2009, Onondaga County announced the formation of a Fisheries Advisory Board. Nominally, the board's mission was to advise the county on "lakes and streams and their growing popularity for fishing." But the main target seemed to be Onondaga Lake and its tributaries and the impacts of cleanup operations on those waterways. Five local sportsmen, all primarily from the area's fishing community, were included on the board, as well as a hatchery aide and a local sportswriter.[26]

Save the Rain

The resolution to the Midland Street sewage facility, one of the truly bright spots in the cleanup effort, was approved by Federal District Judge Frederick Scullin, in November 2009. When Joanie Mahoney had assumed office as the first woman county executive in Onondaga County history, the Atlantic Legal Foundation had asked for a temporary reprieve from the Court's order to move forward with the final phase of the Midland plant, to bring the County into compliance with the 1998 Consent Decree. The judge had agreed to a one-year moratorium on further constructions of all CSO projects while the County began implementation of various green infrastructure projects. By the time of Scullin's current order, the County had made some progress via its Save the Rain program, completing five projects, projecting completion of three more by 2011, and having ten more in the pipeline for approval.[27]

Green infrastructure projects are designed to slow the flow of rainwater and snow melt into sewer systems. In cities, with acres of concrete sidewalks and asphalt streets, rain runs quickly into sewer lines, where it is funneled and flushed into treatment plants, streams, and rivers, and can even end up in residential basements. In combined sewage systems, it becomes mixed with toilet flushings. Multiple strategies can be implemented to prevent a system from being overwhelmed. Porous pavement, pervious concrete, and infiltration trenches (which, as the name implies, collect runoff and allow it to seep into the soil) can slow runoff from hard surfaces. Rain gardens, cisterns, swales, planters, rain barrels, and green roofs also have the effect of slowing runoff in major precipitation events. There are multiple advantages to a green infrastructure approach.

Rain gardens, planters, and green roofs, provide additional greenery that can, like city parks, and tree plantings, mitigate heat island effects endemic to cities large and small (especially useful in the age of climate change). They also remove pollutants from rainwater, sequestering it biologically before it reaches waterways. Given their public nature, green infrastructure projects can remind residents of the value and necessity of a city's public health infrastructure, so often hidden from view. Such

projects can add aesthetic and economic value to a neighborhood, compared to large industrial sewage facilities with their multiple attendant
negative impacts. Green infrastructure is much less expensive than traditional gray approaches to stormwater management.[28] It can have impacts on an entire community's sense of self-esteem.

In the case of Onondaga County, the Save the Rain program changed
its reputation from a place engaged in environmental injustice against
an historically Black neighborhood, to one gaining national prominence
for its innovative green initiatives.[29] The County eventually completed
more than 200 projects, dozens of which are featured on the Save the
Rain website.[30] As such, the Syracuse metro area has become a model
for how cities with aging sewer infrastructure can retool themselves for
a greener future. Joanie Mahoney would eventually become President
of the prestigious State University of New York Environmental Sciences
and Forestry school. But while mitigation of Onondaga's Lake's sewage
problems was turning into a major success story, industrial contamination remained and proved to involve a much more intractable set
of problems, for which no viable green infrastructures solutions were
available.

The Dredging Plan: Would You Want This in Your Backyard?

In January 2010, the state DEC finally began taking comments and
suggestions regarding Honeywell's plan to dredge the lake bottom and
relocate the sediment to the LCP waste site, known as "Allied Wastebed
13" in Camillus. Officials were also looking for comments on Honeywell's "habitat restoration plan" to be developed for the southern end of
the lake.[31]

One hundred people attended a public hearing, which consisted primarily of angry outbursts from residents of Camillus, who objected to
the dumping of lake bottom sediments in their community. State DEC
Director Ken Lynch took the brunt of the criticism. When asked by one
resident, Jo Personte, whether Lynch would want such a facility in his
community, he responded, "There is no one that would like such a facility in their backyard." But, he noted, that residue had to go somewhere,
and the 161-acre waste facility was, according to him, "safe" and would
be "largely unnoticed." Lynch contended that a number of precautions

had been taken into account to limit the impact of the waste removal. The waste would be pumped through a pipe into 100-foot long by thirty-inch diameter tubes made of permeable plastic. This would eliminate the need for truck transport. Wastewater would then leach out, be treated at the Metropolitan Sewage Treatment plant, and be returned to the lake. Of course, the still highly toxic residue would remain buried at the site indefinitely. But this process, Lynch contended, would limit disruptions of traffic and noise as well as minimize potential problems from odors. The plan was still only in draft form, but, if approved, the dredge would begin in 2011.[32]

A letter to the editor that followed a week later signaled further evidence of local discontent. It expressed dismay that the dredged material would be located not only close to a residential area, but also to Solvay Middle School.[33] Two weeks later, the town board voted 7–0 to keep the waste materials out. Board members were especially concerned about the three new housing developments that had been constructed near the proposed dump site. Town Supervisor Mary Ann Coogan suggested that the town had the legal authority to keep the plan from moving forward. "Our attorney firmly believes we can tell them 'no,'" she stated.[34]

One resident, George Mezey, proposed an alternative idea, which was to put the material behind a steel wall built between Harbor and Ley Creeks. This, he contended, would have the added benefit of creating a park along the creek. As a project manager involved with building the new sewage treatment plant and some aspects of the Destiny USA mall in Syracuse, Mezey had some expertise. One admitted drawback for his plan is that it would fill in part of the lake. DEC Director Lynch responded that it would be difficult to get approval for a plan that actually left the sediments in or very near the lake itself.[35] In the end, Mezey's proposal went nowhere.

Residents continued writing letters to the *Post-Standard* expressing concerns. One worried about the town's image if the project went through. Another noted the potential for escaping sewer gas because of the raw sewage that was at one time dumped on the lake bottom. The DEC was perceived as heavy-handed and insensitive to residents' concerns.[36] Its lack of transparency also came under criticism.[37]

When the new mayor of the Village of Camillus entered office in April, he took aim at the proposed waste site. Mike Montero, a violinist, music store owner, and transplant from Staten Island, was adamant, "I think everybody should know about this, about what it's going to do to property values and join the fight to make it go somewhere else. Everyone who lives in Camillus—and even towns bordering it—should be concerned."[38]

When the DEC released a twenty-two-page report compiling information related to the plan, residents, including town Councilor Mark Kolinks, called it "propaganda." Tom Gdula, co-chair of the Camillus Community Coalition said, "There's nothing new in there . . . It is the same marketing propaganda that does not provide the informed person with any level of comfort that the real concerns for safety and long-term health are being addressed. I don't know if we've moved any further down the road."[39]

In early spring 2010, Honeywell, amid some fanfare, planted native grasses and shrubs on a portion of the wastebeds between the lake and Interstate 690. A company spokesperson, Victoria Streitfeld, predicted that the area would soon be awash in flowers, amid the shrubs and grasses. But by May, nothing had happened. Neither Honeywell, nor the firm that did the actual planting, was willing to discuss the matter. Various theories were put forward. One was that the plants simply did not get enough water to become established. Others focused on a potentially more sinister cause. Tom Gdula believed that the calcium compounds were so contaminated with mercury and other materials that they simply could not sustain life. "It would be a miracle that those mounds are safe and contain only calcium carbonate," he stated. Honeywell, for its part, noted that the mounds were shaped to mirror glacial moraines, so ubiquitous in the upstate New York landscape.[40]

In spite of the setback, Honeywell was now using the Onondaga Lake cleanup to burnish its corporate image. A series of advertisements in the *Post-Standard* were part of the effort. One highlighted the company's collaboration with Audubon to establish an Important Bird Area program for the lake and its watershed. The ad, drawing upon support from the New York State Department of Environmental Conservation, noted sightings of bald eagles and peregrine falcons. The plan established eleven species of birds slated for the restored habitat. The ad stated, "Bird habitat and birding are important to the future of Onondaga Lake, and Honeywell is partnering with Audubon in developing educational programs to raise awareness and promote stewardship for future generations."[41]

A second ad featured hydroplaning, crew teams, and the local yacht club. "Along the lake's shores," it contended, "more than 1.3 million visitors annually enjoy the network of hiking and biking trails, the region's premier skate park, waterfront recreational ball fields for softball, bocce, shuffleboard and volleyball, plus quiet tree-shaded areas for family picnics and relaxation." While the ad noted that it would take a number of years before the lake was restored, in the meantime, "thousands already agree that Onondaga Lake is their favorite place to be for on-the-water fun and outdoor recreation."[42] A third ad, titled "Attracting Wildlife," acclaimed the shrub willow area, calling attention to its potential to increase biodiversity and to provide biofuels.[43]

Despite Honeywell's efforts, some area residents were unhappy with the waste disposal plan. Camillus residents had asked for a health assessment from the federal Environmental Protection Agency regarding the potential impacts of the proposed project. In response, the EPA minimized any potential health problems, leaving opponents dissatisfied. One alternative being floated was to apply "chemical oxidation" to the waste to eliminate potential threats. The plan was not without its proponents. Rick Hevier, who was involved with the free market-oriented group Facts not Fear, dismissed claims of mercury and PCB contamination of the dredged material, contending that it was essentially harmless.[44]

In June 2010, the New York State Department of Health rescinded an earlier finding that it was okay to eat some of the fish in Onondaga Lake.[45] In 1970, the state put a complete ban on eating the lake's fish into effect, but in 1999, it approved consumption of some species, a finding that was again reconfirmed in 2007. Nevertheless, recent studies suggested that levels of PCBs and mercury in all but the lake's smallest fish (bullheads and pumpkinseeds) were unhealthy. Not everyone had agreed with the 1999 change, including Syracuse University Professor Charles Driscoll, an expert on mercury contamination: "I thought then it was a mistake, and it's clear now that it was." Health Department spokesperson Claire Pospisil refused to back away from the 1999 decision, insisting that the changed recommendation resulted from new data. The state contended that contaminant levels in the fish, PCBs in particular, had risen over time, although it was not clear why. Regional DEC director Ken Lynch suggested that as the lake became cleaner, fish were moving into areas that had not previously been habitable, but now were, and these areas still contained high levels of contamination. The DEC seemed reluctant to admit that there might be new, but not entirely understood, sources of contamination, or that their original analysis might have been wrong.

That same month, the EPA released its report supporting the Camillus burial site, known as "Wastebed 13," contending that it posed no significant health risks to local residents. "All resulting estimated risks were within levels identified by the EPA as acceptable," the report noted.[46] As might be expected, it failed to convince residents. A letter to the editor of *The Post-Standard* summarized the concern. Resident Alex Abda criticized the EPA for basing its findings on the proviso that Honeywell's plan would work exactly as predicted by its engineers. Independent engineers working with the opponents, on the other hand, concluded that "*the current plan is rife with flaws and likely to fail in a number of respects*" (emphasis included). Moreover, Abda criticized the plan to establish spill contingencies and argued that there was inadequate monitoring.[47]

Honeywell was not waiting for acquiescence on the part of town residents. In fact, it placed a request with the DEC to have it exempted from local permitting requirements. Kenneth Lynch stated that such exemptions were routine "but not automatic." The state would take Honeywell's requests seriously, given that its proposal fell under the purview of state and federal jurisdiction that could override local ordinances. Still, certain criteria had to be met, for instance, that waste processing or remediation would have to occur at the disposal site and that requirements for public hearings and comments had to be met.[48]

It didn't take long for the state to act. On August 10, the DEC granted Honeywell's request for exemption from the local permitting process, although it would have to comply with the "technical requirements" of existing permit rules, which primarily involved meeting zoning ordinances. Honeywell would also have to pay the fees that would have been required had it been required to apply for permits.[49]

By July, the mounds that Honeywell had created from the excavated soil near the LCP facility were starting to turn green. Their previous incapacity to do so appeared to be the result of a lack of rain. but June brought more than six inches of precipitation, and the rye grass planted on the mounds responded accordingly.[50]

Around the same time, Congressman Dan Maffei announced his introduction of the "Onondaga Lake Restoration Act," which would turn control of the cleanup over to local groups, rather than having it directed entirely by the Army Corps of Engineers. The EPA would be the federal agency responsible for overseeing the effort, but it would be required to consult with representatives from the Onondaga Nation and various other local constituencies. Senator Kirsten Gillibrand introduced a similar bill in the Senate. The legislation would open up a research center on the lakeshore where scientists from Syracuse University, the SUNY Environmental Science and Forestry school, the Upstate Freshwater Institute, and the Onondaga Environmental Institute would be able to collaborate on research and monitoring activities. It would also establish an Onondaga Lake Watershed Council, representatives of which would oversee the management of the cleanup project.

Local environmentalists were highly supportive of the legislation. Steve Effler of the Upstate Freshwater Institute in Syracuse said, "I think it's great . . . I see nothing but good coming out of this." Dereth Glance, executive program director of Citizens Campaign for the Environment, stated, "We're very excited about the legislation." Stakeholders in the cleanup had been particularly dismayed that decisions were being made from the Buffalo office of the Corps of Engineers, without involvement of local people. This legislation would change that, but not everyone was pleased with it. Republican Congressman James Walsh argued that Honeywell should have a seat at the table, given that it was investing nearly $500 million in the project. He considered it a "glaring omission" that they didn't have one. Moreover, he wanted to add voices from the town of Camillus, where the waste was being buried.[51]

Steve Effler, on the other hand, was particularly pleased with the turn of developments that the Maffei legislation represented. Effler had spent much of his life committed to a lake cleanup. He was thought to be the person who dubbed Onondaga "the most polluted lake in America." As founder of the Upstate Freshwater Institute, he had strongly criticized early cleanup proposals, which he considered inadequate. As a result, the Institute had become anathema to public officials, resulting in the

loss of "hundreds of thousands of dollars" of potential government funding. Nevertheless, Effler's tenacity was being vindicated. The upgrade of
the Metropolitan Syracuse Water Treatment Plant, which he along with
Samuel Sage had relentlessly pushed for, was finally completed, resulting
in significant improvements in the lake's water quality. Effler had been
an outsider, banging on the doors of public officials and Honeywell executives, but now he was being invited into the process in a very public
way by the Maffei legislation. It seemed as though things were moving
in the right direction, and Effler predicted that the lake would be swim-
mable by 2018.[52]

The controversy around the Camillus wastebeds refused to abate,
however. At an EPA-sponsored public forum, Mark Tracy, with the
support of a hundred other local residents, asked the agency to consider using oxidation and nanotechnology to reduce the toxicity of the
waste before burying it. At the very least, Tracy felt as though the agency
should support a study, at a cost of $200,000, to see if it might work. The
study simply involved putting some of the dredge material into buckets
and sending them to some of the companies using the technologies to
see if they could be successfully applied, but the EPA showed little interest in Tracy's proposal. EPA project manager Robert Nunes questioned
the necessity of a study, much less the application of this technology,
given that contamination levels in the calcium chloride were already
very low.[53]

In the meantime, work continued to prepare the wastebeds to receive
their contents. Workers began to put clean fill at the bottom of the "old
dumping area" near the Syracuse Airport, where the water treatment
plant would be located. Nearly 90% of the waste was water, and it would
be treated at the facility and returned to the lake. The remaining sediment, mostly calcium chloride, would remain at the site. Opponents to
the plan contended that the remaining materials were contaminated with
mercury and PCBs, but the state continued to minimize the amounts
involved and the risks associated with them. The construction of the
pipeline to transport the wastes was slated to begin in spring 2012.[54]

In November, Honeywell came to an agreement with the state DEC
to plant willows on 670 acres of the old wastebeds, the first step toward
making the area into a park. The project would also involve mechanisms
to control the mudboils and included lakefront improvements, such as

a new dock. The project was estimated to cost approximately $2 million. The willow-planting project was considered to be a "green" alternative to the usual practice of hauling in tons of fill to cover the waste and then planting over it. The willows were touted as a natural cap. Once established, they would have the advantage of reducing the entry of water into Geddes Creek and Onondaga Lake, due to their capacity to absorb water. As part of the overall project, which was expected to take ten years to complete, Honeywell also agreed to open its lake shoreline property to public fishing.[55]

Not everyone was convinced that tree planting scheme was the best alternative for dealing with the massive piles of calcium chloride. Clyde Ohl, writing to the *Post-Standard*, accused Honeywell, with the state's collusion, of taking the cheap way out. Ohl noted that the standard way to cap an industrial waste site was to cover it with three feet of topsoil. If that were done in this case, it would cost about $120 million. According to Ohl, "Allied-Honeywell effectively lobbied our too-timid local and state leaders to defer and ultimately eliminate covering the wastebeds." He continued, "Allied-Honeywell's goal is to resolve the wastebed issues and get out of town at the lowest possible cost."[56]

In April 2010, the Onondaga Nation released its "Vision for a Clean Onondaga Lake," which was highly critical of the state-sanctioned Honeywell plan. The report was especially critical of the proposed lake-bed cap and the limited area to subjected to dredging. The Nation's approach, rather than focusing on recreation and tourism, emphasized the sacred nature of the lake. Onondaga Chief Jake Edwards argued that "we can set precedent and have a glass of water from the lake that we can drink and show the world what we can do when we work together."[57] As things moved forward, the Nation's vision of the lake would be ignored within official circles.

In 2011, Rochester developer Thomas Wilmot was proposing to develop a racetrack and gambling facility, known as a "racino" in Onondaga County. One potential site was partially on land that had formerly been an Allied Wastebed and was now being designated as a parkland and recreational area. As the *Post-Standard* noted, the cleaning up of Onondaga Lake was creating new pressures for development along its shorelines.[58] The potential economic value of a cleaner lake was becoming increasingly apparent.

In March 2011, Citizens Campaign for the Environment took out a full-page ad in the *Post-Standard* to announce that the dredging of Onondaga Lake would begin in 2012. The ad included a list of other toxic sites on the lake and plans to deal with them as well: National Grid was cleaning up its old gas plant. General Motors was in negotiations with regard to PCB contamination in Ley Creek. The advertisement said, "You have a stake in Onondaga Lake. Onondaga Lake is our lake; it defines our community's history and our shared future."[59]

Proposed legislation to turn local control of the lake cleanup from Honeywell over to control by local community groups gathered further momentum when Representative James Walsh decided to support it. Walsh's chief objection was that Honeywell had been excluded from the panel that would oversee the cleanup, but a revised version of the bill changed that. Questions remained, however, about whether the new Republican congressional representative from the district, Ann Marie Buerkle, who replaced Dan Maffei, would support the bill.[60] The exact fate of the bill remains unclear, but the cleanup effort was never turned over to local organizations. Had it been successful, it would have expanded citizen participation in the cleanup operation, and would have included representatives of the Onondaga Nation.

"What's with the bass fishing in Onondaga Lake?" asked *Post-Standard* outdoors editor David Figura. The lake had recently become home to several nationally recognized bass fishing tournaments. Up through 2007, Figura noted, it was one of the nation's "best kept secrets" for bass fishing. But something had happened, and the bass population seemed to have gone into serious decline. In June, a local competition came up empty, without a single bass being caught. One angler reported seeing fish but was not able to attract them to a lure, leading to speculation that the bass were still in the lake, though for some reason no longer tempted by bait. Lars Rudstam, a Cornell University biologist, noted that the lake was full of alewives, a fish that bass feed upon. As a result, he speculated, the fish simply weren't hungry enough to be tempted by artificial lures. Neil Ringler, a SUNY ESF biologist, suggested that the bass had moved to different parts of the lake, in response to changes in the distribution of aquatic plants, while anglers had not kept up with the change. Both researchers were in agreement that there were indeed bass in the lake, some as large as seven pounds.[61]

As the lake was prepared for dredging, researchers began to take a re-newed interest in what might be on the lake's bottom. An archeological study revealed a virtual treasure trove of local history. Researchers dis-covered eight sunken boats, including a tug barge, the Stillwater, which was hauled out to the middle of the lake and scuttled in 1940 after being damaged by ice and declared unfit for use. They also found a jet plane that had crashed into the lake in 1955. Sonar revealed remnants of old docks, piers, and pipelines, fragments of the numerous resorts that once lined the shore before being driven out of business by the lake's con-tamination in the 1930s. The iceboat Blitz, which had been involved in a head-on collision with another iceboat in 1904, was also discovered. Two operators had died, and the boat was left on the lake and allowed to sink to the bottom when the ice melted in the spring.[62] These discoveries are indicative of the careless attitude, if not outright contempt, that city and county officials towards the lake's condition in an earlier period. All kinds of materials either fell into or were dumped into the lake, with no attempt to find or remove them.

Dredging Begins

In spring of 2012, dredging operations finally began. The largest of three dredges, designated as "Marlin," was a huge piece of machinery, con-structed in Louisiana for the specific purpose of dredging Onondaga Lake. The dredge was shipped up in pieces on eleven trucks. The Marlin dredge was forty feet wide and 100 feet long, and once up and running, it would operate twenty-four hours a day, drilling into the lake bottom to break up the muck and then pumping two million yards of contaminated sediment into a four-mile-long pipe, from which it would be deposited in the controversial dump site in Camillus. There, the sediment would be placed in large plastic geotubes, allowing the water to drain out. The water would then be treated in two different facilities and pumped back into the lake.[63] Once dredging of the 215 most contaminated acres of the lake bottom was completed, along with 235 acres of less contaminated sediment, the dump site would be capped with sand and gravel, to com-plete the nearly half-billion-dollar operation.[64] However, as had been the case from the beginning, not everyone was convinced that the project was adequate. Attorney Joe Heath, representing the Onondaga Nation, contended that "A great deal of toxins are going to be left in the lake. We think this is going to come back and haunt the next generation." The cost of dredging the entire lake bottom had in 2005 been estimated at more than two billion dollars. It was clearly in Honeywell's interest to accept the less extensive dredging approach.[65] Marlin began operating in August. It would take four years, only stopping when the weather became too cold for the dredge to work effectively. The capping operated simultaneously with the dredging. As material was removed, sand and gravel were spread over the site.[66]

In September, County Executive Joanie Mahoney made the cover of the trade journal *Municipal Sewer and Water*. Mahoney was being celebrated for the now nationally recognized Save the Rain initiative, designed to keep storm runoff from entering into the Onondaga Lake watershed. The innovative project used green infrastructure to capture rainwater before it could enter into the storm drain system. Absorbent roofs, porous pavements, rain gardens, bioswales, and cisterns were being installed throughout the county, with the hope of capturing 43.6 million gallons of stormwater. In addition, the county was planting 1,000 trees, which would capture another two million gallons of runoff. As a result, the county would save $20 million by not having to build storage plants to capture the runoff before diverting it into the municipal sewage facility. As the program moved forward, the lake was declared by Onondaga Lake Partnership to be the cleanest it had been in 100 years. Nearly two hundred species of birds had been detected, including herons, eagles, and ospreys. Plant cover was increasing at a fast rate as well.[67]

But the dredging operations, in spite of DEC and EPA assurances, were creating serious problems for the residents of Camillus. The obnoxious smells being generated were so bad that operations had to be halted. One resident, Lynda Wade, described them as "Really bad, flammable, chemical, eye-tearing, choking, can't-stay-outside-in it, can't breathe odors." The air smelled like mothballs and body odor. Ken Lynch, regional DEC director, was, as usual, dismissive of the problem. The air was being monitored, he stated, and nothing hazardous had been detected. The mothball smell was simply naphthalene. Residents were not mollified, and Town Supervisor Mary Ann Coogan suggested that unless the odor issue was addressed there might be an "uprising" by local residents. She also threatened a lawsuit. The smell had started four months after the dredging began, and Honeywell claimed it was being responsive. It promised to dredge the gunk more quickly and install misting and carbon filtration systems to suppress the odors.[68]

I spoke with Lynda Wade about her experiences with the dredging. This was in 2021, nearly ten years after the operation had started, and seven after it was completed. We spoke in her open garage, on a bright summer's day, in a tidy suburban enclave, in the town of Camillus, New York. Although it had been a number of years since the material had been deposited a few hundred yards from her home, the trauma seemed fresh in her mind. She still had folders related to the project, with news clippings, press releases, legal documents, and other materials related to her role as one of the leaders of the Camillus Clean Air Coalition. She is, to say, the least, not a fan of the Honeywell, the DEC, or the USEPA, in terms of how they handled the situation.

Wade's house sits at the bottom of a hill, on the top of which was Wastebed 13, so the ambient substances would descend down the hill into a dip just below the back of her house, where they would collect and then make it into her neighborhood. Wade told me that the smells were intense and varied. They made her wretch and sometimes vomit. The mothball smell, naphthalene, a recognized toxic, was bad, but the body odor smell was the worst. The odors began with the dredging in early spring and continued until frost for three years. She and her husband had purchased the house recently, without knowing that the dredging was in the works. Wade noted that they were able to get an injunction against the dumping operation for ten days after it began, at which point Honeywell put up some misters that they and the DEC claimed would mitigate the problem. They didn't. One problem was the intermittent character of the smells. Wade would call the DEC to send someone out, but by the time they arrived, a particular episode might have abated.

In order to mount a legal challenge to the operation, and to get a sense of what the nature of the toxics they were being exposed to involved, the Camillus Clear Air Coalition hired the New Jersey environmental consulting firm, Minnich and Scotto. I spoke to Tim Minnich, and he supported the charges made by Lynda Wade. He was very critical of the way that Honeywell, the DEC, and the USEPA dealt with the dredging. He said that the slurry of material from the lake, because it had to be pumped three miles uphill, was under intense pressure, so that when it was released into the air, it created a large plume, a kind of geyser, of toxic mist. The misters that Honeywell put up to address this were useless. Minnich also contended that the DEC abrogated its own rules, because it never conducted a pilot study, required in the Superfund Act (CERCLA) when a site cleanup was proposed. Government environmental agencies essentially created a new Superfund site with the dredge materials in violation of state and national environmental statues.

Minnich also said that the geotubes used to funnel the wastes into were not intended for that purpose but were manufactured (by Honeywell) to stem beach erosion. (This is true, but the geotubes are also now sold as a dewatering technology and for other environmental remediation purposes; whether they were before the Onondaga Lake dredge is not clear.)[69]

Public officials and Honeywell personnel continued to minimize the problem. John McAuliffe, a Syracuse Honeywell official, stated that, while a "nuisance," the odors presented no health threat, saying "I think that all along we expected that we would have fleeting odors." Mike Salvagno, a technician for O'Brien and Gere, the engineering firm in charge of the project for Honeywell, contended that naphthalene had a low odor threshold, so even at levels below acceptable safety standards, its smell was noticeable. The company proceeded with its plan to put plastic tarps over the geotubes into which the waste was being pumped and set up a system of misters in order to further control the odors.[70] Residents did not, however, find these measures to be effective. The odors they experienced persisted.

In October, the next phase of a bicycle/hiking path was announced. The 2.5-mile path would extend the trail farther down along the western shore of the lake, across Nine Mile Creek, and then up a sixty-foot-high mound created from calcium carbonate deposits from the Solvay wastebeds. The artificial hill would be high enough to provide users with a nice view of the lake and city below. Part of the long-term plan was to remediate the tons of residue, but in the meantime, it provided a benefit to those seeking a refuge from the buzz of the city and the network of highways and shopping malls that skirted the lake.[71] The first lake resort, which initiated Onondaga's "Golden Age of Resorts" had once flourished on the very ground now covered by these massive calcium chloride deposits. It was shuttered in 1913, as the Solvay process waste and raw sewage made fishing, swimming, and enjoying the shoreline untenable.[72]

As the lake became cleaner, questions of development emerged. Ironically, the lake's pollution had been such a deterrent to lakefront development that the lakeshore appeared to be in a relatively untouched or natural state. Natural-looking shorelines tend to be magnets for development projects, so as the cleanup moved forward, it seemed increasingly unlikely that the lake's undeveloped shoreline would remain so indefinitely. And how would the improved lake's health figure into this process of change?

By the end of 2012, state officials were suggesting that swimming in parts of the lake was safe, although it was still prohibited. But, even if the lake was deemed safe by state officials, would people actually swim in it? Older residents might never be willing to swim in Onondaga; younger ones, less familiar with its history, might be more willing to take a chance. But being drawn to the lakeshore seemed to transcend generations, hence the interest in hiking and biking paths around the lake. According to David Reed, director of FOCUS, visitors "want to get close to the lake, but that's close enough."[73] Contamination issues slowed the progress of constructing lakeshore paths. Soil had to be removed from some areas underneath and replaced with clean fill.[74]

By February 2013, the mothball stink had returned to Camillus, and residents were understandably upset. As some noted, when Honeywell proposed the project, officials for the company stated that residents of Camillus wouldn't even be aware that it was going forward. But that was clearly not the case, as many complained of odors that burned their noses and gave them headaches. DEC engineer Mary Jane Peachey at least admitted that a problem existed. "We are here," she said at a public forum, "because we believe there is more that needs to be done on multiple levels. This is a priority for us. We are working on this every day. If you as a community have ideas, we welcome them."[75]

An angry letter to the *Post-Dispatch* in March took the DEC and Honeywell to task for the continued smell. Kenneth Lynch was criticized for being "in lockstep" with Honeywell by the Camillus Clean Air Coalition, which stated that benzene levels in the neighborhood had exceeded recommended standards. Lynch came under particularly withering criticism for his suggestion at a public meeting that he regretted that he could not convince area residents that there were no health concerns. The letter responded, "You cannot 'convince' the parents of the children who had more asthma attacks last fall than any other time in their lives. You cannot 'convince' those who suffered nausea, headaches, burning throats and nasal passages, tearing eyes or allergic reactions. You cannot 'convince' those who cannot sell their homes or worried their investments might disappear. You cannot 'convince' the mothers who live in fear and have lain awake at night crying because the very people charged with our safety do nothing. If you fixed the problem, there would be no need to 'convince' anyone of anything." The coalition then took legal action to try to force the company either to improve the situation or stop the dredging.[76]

The firm hired by the town residents issued a report in April that concluded the state had significantly underestimated health risks associated with the disposal site and that it was being improperly monitored. The reported concluded that the mitigation measures put into place by Honeywell, which involved the tarps and misters (previously mentioned), had little impact. The misters exacerbated the problem, because the detergent used in them, while not toxic, was an irritant, creating "yet another airborne compound about which to be concerned." The manufacturer of the substance made no claim that it would have an impact on toxic emissions.[77] Minnich and Scotto contended that the naphthalene exposure had long-term health consequences and that it and benzene exposure created very real short-term negative health impacts. They questioned the air monitoring program, both in terms its coverage and the type of monitoring used.

The report contended that the standard EPA canister monitoring technique, TO 15, underestimated the presence of naphthalene, and recommended the use of TO 16 monitoring. Also at issue was the Human Health Risk Assessment, required to be completed by the EPA before the project could be approved. Minnich and Scotto questioned its adequacy. They noted that dispersion modeling was never completed: "instead of modeling the Facility emission rates, EPA simply assumed that the concentration of each COC [contaminant of concern] along the [site] perimeter would be 'fixed' at the safe-level criteria. They then applied dispersion modeling relationships to dilute these concentrations before reaching nearby residents." In other words, the EPA began with the premise the ambient air at the site was at safe levels, and then used this premise to determine that residential exposure was even safer.[78]

Scotto and Minnich then conducted their own analysis, basing it upon measurements of concentrations of contaminants in the dredge material (publicly available information) and modeling of air currents using meteorological data, and the resultant dispersion of site discharges. They concluded that residents faced acute exposure to three toxic chemicals, naphthalene, 1,4 dichlorobenzene (a chlorine/benzene compound), and benzene, at levels significantly above those determined to be safe by the EPA.[79]

The Scotto and Minnich analysis led to an exchange between the EPA and the firm. The resulting back-and-forth went deep into the weeds of technical aspects of testing protocols, unsafe contaminant exposure levels, and atmospheric dispersion models. While the public authorities apparently felt that the criticisms were serious enough to merit a response, in the end, they rejected any criticism of the project, and pushed forward without accepting any of Scotto and Minnich's recommendations.[80] Residents believed that the Honeywell and the DEC were moving quickly to avoid an injunction ordered by the state court, the potential result of the residents' lawsuit.

The Camillus group faced a setback when Syracuse Judge Frederick Scullin not only dismissed the case but then suggested that the federal Northern District Court of New York should decide whether the plaintiff's Manhattan lawyer, Kenneth McCallion, ought to have his law license revoked. Scullin was furious that McCallion had cited a case that had been overturned in his brief. Moreover, McCallion was thirty-nine days "too late" in filing one motion and had provided "incorrect legal advice" to the Camillus Clean Air Coalition. McCallion's excuse that he had simply made a mistake on citing precedent was "less than convincing to the judge."[81] Tim Minnich characterized the scene in the courtroom as David versus Goliath: McCallion versus the bank of corporate attorneys representing Honeywell. But, of course, in the biblical story, David won, and, in this case, McCallion didn't. In the end, the judge would not allow the Minnich and Scotto report into evidence. The Camillus residents had essentially no chance of getting an injunction against the project when faced with the combined forces of Honeywell, the DEC, and the USEPA.[82] The dredging and depositing would continue for three years, with odors and fears of what they represented becoming an unwelcome part of life in the suburban community from spring until the first frost in early November.

Chemical Remediation

While the dredging operation was facing opposition, researchers had devised a new process to reduce contamination problems within the lake itself. Described as a "homegrown, cobbled together experiment to cut mercury levels" the process was deemed by scientists from the Upstate Freshwater Institute, who had devised it, as a "stunning success." They had experimented with putting calcium nitrate, a cheap and readily available fertilizer, into the water. The fertilizer generated bacteria that in turn trapped mercury from the lake bottom, preventing it from entering into the water or the food chain. The process was so successful that it had generated published research and international attention. Environmental consultant David Austin went so far as to state that "I think it will stand out as one of the great environmental engineering projects of our time."

The decision to use calcium nitrate stemmed from improvements in the Metropolitan Sewage Treatment facility. Researchers observed that as the facility became more effective, ammonia levels in the lake declined, more benign nitrates replaced it, and mercury levels fell. Two scientists from the Freshwater Institute, Dave Matthews and Steve Effler, made the connection and decided to try putting larger amounts of calcium nitrates in the lake bottom. They believed that the nitrates would allow some bacteria to thrive and replace the bacteria that had been converting the lake bottom's mercury into methylmercury, the form that made its way into the food chain. Matthews and Effler concocted a system chaining together twenty-foot pipes to pump the calcium nitrate to three sites in the deepest part of the lake. When they did this, they found that mercury levels dropped at the lake bottom by 94%. It would take time to see how this affected the entire food chain, but the researchers were optimistic. If the process turned out to be as successful as they hoped, the lake's thriving fish population might someday be available for human consumption.[83]

The experiment was not, however, without its critics. Tadodaho Sid Hill, Chief of the Onondaga Nation, criticized a *Post-Standard* article about the project as over-stating its success. He also criticized the research itself, suggesting that the scientists involved had picked a baseline for comparison, 2009, which had especially high mercury levels, thus making the recent reductions appear to be higher than they might actually be. In Tadodaho Hill's view, there was simply not enough evidence yet to demonstrate the effectiveness of the nitrate application, so the jury was still out. Perhaps more importantly, however, Hill noted that the nitrate application would have to be continuous. If it stopped, mercury levels would again rise to previous levels. Honeywell had not yet committed to continuing the program on what would need to be an ongoing basis. What Tadodaho Hill and the Onondaga Nation continued to demand was that Honeywell completely remove the mercury from the sediments in the lake bottom at an estimated cost of $2.157 billion.[84]

The Freshwater Institute Scientists felt the need to respond to Tadodaho Hill. They contended that the decision to use 2009 as a baseline was legitimate because 2010 was uncharacteristically high due to low runoff into the lake from area streams. Mercury levels were indeed variable from year-to-year, and there had been long-term declines, but the scientists held to their position that their experiment was a "stunning success."[85] Their response did not, however, address Todadaho Hill's claim that nitrates would have to be continuously applied for mercury levels to remain lower. This has continued to be controversial and a sore point with critics of the cleanup operation. When I talked to Matthews, he admitted that there was disagreement on this issue. He stated that the process would have to continue for "decades," but he didn't see a practical alternative.[86] Charles Driscoll, on the other hand, was much more upbeat. He stated that it frustrated him when Honeywell agreed to continue with the applications indefinitely, whereas, in his view, due to natural processes of sedimentation, along with the removal of large amounts of contaminated material, nitrate applications could very well stop in ten or fifteen years. Further exploration of these conflicting viewpoints toward the lake cleanup process will be explored in more detail in this book's conclusion.

Multiple Perceptions and Practices as the Clean-up Continues

Honeywell's continuing charm offensive was on full display in its funding of a photography exhibit "Birds of Onondaga Lake." The exhibit, hosted by the Audubon Society, was held at the Onondaga Lake Visitors Center (funded by Honeywell), and included the works of several area photographers. The executive director of Audubon New York Erin Crotty was effusive. "These gorgeous images capture the community's enthusiasm and tell the visual story of why Onondaga Lake has achieved a listing as an Important Bird Area identified by Audubon New York." Even if Honeywell's exhibit was a public relations ploy, there was no denying that Onondaga Lake was home to bald eagles, myriad ducks, and a large array of other waterfowl.[87] Ironically, Honeywell's precursor Allied was partly responsible, since the lake's contamination problems had prevented other commercial and residential development and allowed the bird habitat to thrive undisturbed.

In March 2014, County Executive Joanie Mahoney announced a proposed amphitheater on the shores of Onondaga Lake. It was to be built on a former Superfund site composed of sixty feet of fill that held toxic substances such as benzene, toluene, xyleme, and manganese. But when a hiking/biking trail, which traversed the west side of the lake and came very close to the amphitheater, was completed, the county suggested that a fence be placed around the area to prevent people from wandering into it. Mahoney said that there no plans to remove the contaminated materials, but additional cover would likely be placed on top of it to protect patrons of the new facility.[88]

"County Officials Say: Onondaga Lake Now Cleaner than Some Finger Lakes," heralded the headline of *The Post-Standard* in the spring of 2014. While technically true, it was perhaps a bit deceptive. It referred to only one pollutant: phosphorus. And while it was good news that phosphorus levels in Onondaga Lake had dropped to lower levels than two of the smaller and lesser-known Finger Lakes, Honeoye and Conesus, the lake still had serious contamination problems, especially with regard to mercury. Moreover, Onondaga Lake's phosphorus levels were five times higher than those of Skaneateles Lake, one of the cleaner of the Finger Lakes, which was a protected source of drinking water.[89]

In Spring 2014, Honeywell released an eight-page color brochure touting the progress of its cleanup efforts on Onondaga. Company official John McAulliffe was quoted in it, saying that "Many in the Central New York community are passionate about the cleanup, and there is truly a sense of pride in the progress to date." The company said that the dredging operation was almost complete and that the entire project would be finished by 2016. But members of the Onondaga Nation remained unconvinced. According to Joe Heath, "The state said it would take $2.3 billion and (Honeywell) is spending $450 million. They're crowing about an achievement that's less than 20% of what should be done."[90]

When the West Shore Trail Extension opened in May, an event that involved participation by twenty-five community groups, representatives of the Onondaga were absent. The trail extension tracked over the old Solvay wastebeds, which had not been remediated, but covered with fill instead. The national EPA advised the county to build fences along the trail to keep people from wandering into contaminated areas. Attorney Thane Joyal, speaking for the Onondaga, said, "The nation doesn't see as much to celebrate as other people do . . . the nation isn't really happy about the status of the lake."[91]

Soon Honeywell and the DEC would come under criticism for not releasing detailed reports on the lake cleanup efforts. The *Post-Standard* filed two Freedom of Information requests that were denied by the DEC. The rationale given was that Honeywell was under a consent decree, and a release would "interfere" with ongoing state enforcement actions. But Bob Freeman, executive director of the Committee on Open Government, was not impressed. He asked, "If Honeywell is the subject of the consent order and has the same records as the state investigating body, how could it be contended that disclosure would interfere with the proceeding?"[92] It was a good question.

At the same time, the waste disposal problems in Camillus had hardly been resolved. Kristian Larsen, a lawyer for the environmental group, Civilian Environmental Corps, announced that they would be filing a lawsuit against the EPA for allowing the dredging operation to continue in spite of the serious air contamination problems being generated by it. Larsen stated that at least ten people were "very sick" as a direct result of the dredging. Moreover, Larsen cited the consulting group Minnich and Scotto, which contended that 16,000 pounds of mercury had already been released into the air. Honeywell was dismissive, stating that Larsen's arguments were the same put forward in a lawsuit by the Camillus Clean Air Coalition, which had already been dismissed by a judge in the separate case. Victoria Streitfeld, a spokesperson for Honeywell, said that the project was designed with "the protection of residents in surrounding communities" in mind. The pipe that carried the dredged material, she contended, was thick enough to prevent any contaminants from being released into the atmosphere.[93] In June, the dredging began again.[94]

County Executive Mahoney was also trying to assure residents that the proposed amphitheater would be a safe place to attend events. Deputy Executive Matt Miles, said that it would "probably be as safe as a green field." While the contaminants wouldn't be removed, they would be covered with one to three feet of topsoil. The EPA's assessment seemed less than resounding. It declared the project "acceptable" and recommended that workers wear masks as construction proceeded.[95] In spite of these assurances, Clan Mother Wendy Gonyea told me that members of the Onondaga Nation who worked on the site detected problems. They were required to sign nondisclosure agreements but eventually did talk about their experiences. They said that heavy equipment sank right into the ground during construction. The amphitheater sits on pilings, more than sixty feet long, that are rooted into the bedrock below. While these stabilize the facility, their necessity raises questions about the long-term stability of the areas surrounding it. (She also said that, if Willie Nelson played there, she'd probably buy a ticket.)[96]

Local fishermen continued to be pleased with the quantity and size of fish they were catching in the lake. The DEC had documented sixty-five species of fish in the lake, whereas there had been as few as nine in the 1970s. In addition, the dredging operation seemed to have no appreciable negative effect on the fish populations. Of course, the state was still prohibiting consumption of the fish, but SU's Charlie Driscoll predicted that mercury levels were continuing to fall and that fish would be edible in about ten years.[97]

While pollution problems in the lake were well known, as it became cleaner, biologists began to focus on the less visible problem of invasive species. ESF's Don Leopold stated that central New York had "some of the most serious invasive species in the Northeast, and they're just as bad around the lake as anywhere." Flora included purple loosestrife, which had become ubiquitous along New York State highways, and the fast-growing and prickly European buckthorn. Along with common reeds, these species had pushed out native plants on the shoreline into the lake's shallow waters. In the water itself, round gobies appeared, a species of fish that displaced native species, and which had invaded the Great Lakes and the mid-Atlantic region.[98]

As the dredging and capping operation moved toward conclusion, geologists were focusing more attention on the Tully mudboils. The bluff overlooking the valley was sinking, creating pressure on the underground spaces beneath it. The boils resulted when water running down the hills of the Tully Valley ended up in these spaces, at which point it was pushed to the surface, generating a sea of mud. The mud flowed first into Onondaga Creek, and then eventually into the lake itself. The resulting silt mounds were causing flooding in Syracuse and degrading water quality in both the creek and the lake. Geological evidence suggested that the boils were hundreds, if not thousands, of years old, but starting at the end of the nineteenth century, the brine mining that took place to provide feedstock for the Solvay process was thought to have exacerbated the problem by creating pockets of open space. Clan Mother Wendy Gonyea told me that families used to picnic along the creek. People swam in it, and fish, especially trout, were plentiful. Her brother still fishes in the creek, but Gonyea considers this inadvisable, and especially warned against eating the fish.

To deal with the boils a berm had been constructed—a system of pipes and valves to relieve the underground pressure. This seemed to work for a while, but the problem had returned. The county created a committee charged with finding a new solution. The subsidence from one bluff's collapsing had not only undermined water quality, it had damaged a bridge, a pipeline, and a telephone cable.[99] When I visited the creek at a bridge, where it runs through Onondaga territory, it appeared to be a stream of muddy water. Whatever engineers were doing in an attempt to mitigate the boils effects, it wasn't working, or at least working well enough to clear the creek of mud. And the turbidity was and is still impacting the southern end of Onondaga Lake.

The amphitheater proposal, which had generated a fair amount of controversy primarily because of its cost and doubts about potential profitability, was finally approved by the county legislature in November 2014, when Jamesville representative Linda Ervin changed her vote. She had some doubts about the project, but had concluded that the promise of jobs, especially for women- and minority-owned businesses, made it worth the effort. County Executive Joanie Mahoney's enthusiasm for the project had been crucial. But the state was loaning the county $30 million that would to be paid back at a rate of $2.5 million a year. This meant that approximately 200,000 people would have to attend events to reach the break-even point. Given the number of other event venues in Syracuse, some had their doubts about the feasibility of this proposal. Most vocal was Kevin Holmquist, who suggested the project had been "rammed through."[100]

At almost the same time as the amphitheater's acceptance, John McAuliffe, Honeywell's program director, announced that the lake dredging operations had been completed one year ahead of schedule. Steven Effler was pleased, describing the lake's "renaissance," and its "remarkable" improvement. But Joe Heath, lawyer for the Onondaga, continued to be critical, noting that most of the mercury and other contaminants still resided at the lake bottom, posing a problem for "future generations." The lawsuit brought by Camillus residents to stop the dredging had been unsuccessful and was caught in a kind of legal limbo. The capping process continued, and while the company would have to monitor water quality for years to come, its major contribution to the cleanup effort was now virtually complete.[101]

Still, Camillus residents were not satisfied and filed another lawsuit against the company. The fifty-nine plaintiffs in the case, fifty-one of whom were children, sought damages for the negative health impacts stemming from exposure to toxics during the dredging and disposal process. The state and Honeywell insisted that the process had been completed according to established standards. But residents contended that the monitoring of mercury had been inadequate, in fact, non-existent, despite Honeywell's own analysis that showed the escaping mercury posed a significant problem. So, while the dredging operation may have been completed, the impacts, residents contended, were still being felt.[102] Lynda Wade told me that psychological effects still linger. She explained that when someone is diagnosed with cancer in the neighborhood, especially if they are young, it creates suspicions about the dredging, and the worry extends to potential impacts on your own health and the health of your family. It would be impossible to prove a direct connection, but once the genie is out of the bottle, it's difficult, if not impossible, to get it back in.

Another contaminated site on the lake, the former Roth Steel facility, had been under consideration by Adam Weisman as a possible hotel site. But as it became clear that cleanup costs made the project prohibitive, he switched gears and proposed it as a site for a scrap yard. Opposition formed immediately, since this seemed contrary to efforts to clean up the area. And the controversy surrounding Weisman's proposals deepened when it was alleged that Joanie Mahoney's former chief of staff, Ben Dublin, had been hired (by whom it was not clear) to help facilitate opposition to the projects, an allegation that was never substantiated. A bankruptcy judge had approved sale of the site to American Iron & Metal, a Canadian metal processing firm, but it was doubtful that Roth's permit would allow a scrap yard to be established on the property. Mahoney denied any collusion with outside interests; contending that her opposition resulted from doubts about the compatibility of the scrap yard with current development efforts, she encouraged legal action by the county to try to stop it.[103]

Meanwhile, plans continued to make the lake and the areas surrounding it more accessible to citizens. New York State announced funding for a Loop-to-Lake Trail between the proposed amphitheater and the shoreline, part of a larger plan to connect the trail to a future boat launch site and the City of Syracuse Creekwalk. The boat launch would also include a visitor center, a launch designed specifically for kayaks, and parking.[104]

While the cleanup effort had its skeptics, others embraced it as a major environmental achievement. Audubon New York, for example, gave Honeywell its highest honor, known as the Thomas W. Keesee Jr. Conservation Award. In doing so, the organization stated that the Onondaga cleanup was "one of the most ambitious environmental reclamation projects in the United States."[105]

In July, something of a milestone was reached, when a group of fifty people, featuring notable public figures including County Executive Nick Pirro, legislators Kathy Rapp and Tim Burris, and DEC Commissioner Joseph Martens, jumped from a boat into Onondaga Lake. They remained in the water for approximately ten minutes. An audience, numbering more than one hundred people watched from the shorelines. But a group of twenty Onondaga youth protested from the shore with signs warning of continued contamination and stating, in one case, "Our Land Is Sacred." Nearby residents said that the lake still smelled bad. Even Judith Enck, the EPA's regional administrator, questioned the wisdom of the action, calling it "irresponsible," because it sent the message that the cleanup was completed, while they still had "a long way to go."[106]

In the fall, County Executive Mahoney put forward plans in her budget to spend $300,000 to determine whether a beach should be opened on the lake, and whether swimming could again return on a large scale.[107] But in January 2016, it was reported by Honeywell to the state that the much-vaunted cap, which was now 90% complete, had failed three times since it was first put into place in 2012. The capped material, sitting in a large mound at the lake bottom, apparently came apart, allowing it to slide down to the lower lake bottom, contaminating forty acres of it that had previously been untainted.

The DEC's Ken Lynch and Victoria Streitfeld, a spokesperson for Honeywell, minimized the extent of the problem. Lynch said that the area affected was "small" and that most of the cap would work as advertised, "despite the few disturbances." Streitfeld contended that problems were typical in any "large construction project" and that they would be addressed as they occurred. Sidney Hill and Joseph Heath, Onondaga Nation critics of the project, were in the position of saying, "I told you so." Hill stated, "They had a swimming exhibition out there and they're talking about a clean lake. It's just not true." Heath, the Nation's lawyer, also complained about secrecy related to the cap's failure. Lynch asserted that they released the information in May 2015, as soon as they had it, once they knew "the full extent of these disturbances."[108] This suggested that they knew about the problem "to some extent" previously but had not alerted the public or interested stakeholders for an indeterminate period of time.

Conclusion

An Uncertain Future

In 2013, residents of Camillus, the community which had been the repository of dredged Onondaga Lake sediments, brought a lawsuit against Honeywell for damages due to the impacts on their community. In 2015, a federal district judge in Syracuse denied their claim. Judge Vera Scanlon ruled that since Honeywell had followed both state and federal guidelines regarding the cleanup, the company could not be held liable via additional private lawsuits. Undeterred, however, the residents filed an appeal in the Second Circuit federal appeals court. That appeal was rejected on the same grounds. The residents were caught in a bind. They were asking for damages in state court, but in doing so, they were objecting to the consent decree that had been negotiated between Honeywell, the New York Department of Environmental, and the USEPA. Since the consent decree followed federal and state risk and safety guidelines, Honeywell's actions could not, by definition, in legal terms, have been harmful.

Honeywell had a point. If they had acceded to the state's regulatory authority, then would it be fair to hold them accountable to a potentially different set of standards in a civil jury trial or other legal proceeding? The appeals court (following the district court's judgment) put it as follows:

> CERCLA [The Comprehensive Environmental Response, Compensation, and Liability Act] preempts the residents' claims because the allegations amount to nothing more than a belated challenge to the adequacy of the consent decree itself (as opposed to a failure by Honeywell to comply with or properly implement the consent decree) . . . at bottom the residents were impermissibly arguing, on a state tort law theory, that Honeywell should have departed from the consent decree's terms by conducting additional or different remedial action than the mandate.[1]

The court also accepted the EPA's determination that "'during the period of dredging operations, there has been no evidence of adverse impacts on human health'" and that "'the project is being implemented in a manner which is fully protective of public health.'"[2] Any counterevidence by residents of adverse health impacts was discounted. According to the court, the review process through which the consent decree had been achieved, was rigorous and therefore the cleanup could not be harmful.

The paradox created by this system of environmental regulation transcends the case of Onondaga Lake. Once environmental standards are met by a private entity, citizens have no legal grounds to challenge them. Citizens can challenge their implementation: Did the company follow through with what it agreed to do? (The court ruled that it did in this case.) In theory, citizens have input via public hearings and comments during the review process to address potential gaps in the regulations, but these can be, and often are, minimized or ignored by regulatory officials. Citizens can bring lawsuits to require industry to follow promulgated regulations, but the regulations are not subject to review in individual cases. If they are inadequate or faulty, which is hypothetically possible, citizen complainants must suffer the consequences of those regulations without any chance for legal redress.

In one sense, it is understandable that public agencies and private interests want to avoid endless litigation. Private companies want a degree of finality once they agree to a remediation plan or a set of future practices. But what if citizens have legitimate complaints that have not fully been addressed by negotiated consent decrees, and which may, in some cases, be the result of political pressure or institutional imperatives to achieve a workable compromise? It would seem to be a simple matter of fairness to have their potential harms evaluated in an open process, where they can provide evidence of their claims. Under the current regulatory regime that possibility is ruled out. In Camillus, the residents could smell the bad air. They could still see their kids suffering from asthma and other respiratory issues. In spite of the DEC's claims, residents experienced effects that were not trivial. The problems were discounted by New York's DEC and the federal courts, but a forum for addressing them other than courts, such as mediation perhaps, might be an answer. In the case of Camillus, Honeywell stopped piping dredged material into the community earlier than anticipated. Whether this was the result of technical considerations or political ones remains something of a mystery.

Metro Sewage Treatment

In October 2016, a pipe entering into the Metropolitan Sewage Treatment Plant burst, spilling thirteen million gallons of raw sewage into the creek and Onondaga Lake. The pipe (known as the Ley Creek force main) ran next to some railroad tracks along the southeastern corner of the lake over Onondaga Creek. In 2017, the pipe burst again, resulting in a 4.5-million-gallon spill. Combined spills and leakages had allowed ninety million gallons of sewage to flow into the creek over a ten-year period. Even though the Metro plant had been upgraded, the aging pipe infrastructure was undercutting some of the progress that had been made. The cause of the pipes bursting was pressure from unusual rain events. In 2016, a pipe failed after 6.5 inches of rain fell over a twenty-four-hour period. In 2017, 3.5 inches of rain fell over several days. The county moved to replace the pipe, along with a pipe draining the Liverpool area, at a cost of $18 million.[3]

In 2018, I took a tour of the Metro Sewage Treatment plant. I was included with a group of urban sustainability professionals from Buffalo, Detroit, and Cleveland. We did a walk-through of the treatment process from beginning to end, with detailed explanations by the plant manager.

On entering, a visitor first encounters steel racks that catch large pieces of debris that have been washed into the sewers. This prepares the sewage for a finer sieve-like device, known as the "grit catcher." Early in the twentieth century, Syracuse, like many cities, combined its sewer and runoff systems. As a result, a great deal of detritus gets washed into the system, which needs to be removed. After this, as we walked along, the tour director warned us that it was going to be smelly on parts of the tour. A member of our group remarked that since he grew up on a farm, it wasn't going to bother him. I noticed the smell, but it was tolerable, at least for the short period of time that we were there.

After the grit catcher, the sewage is sent to a well and pump station, from which it is elevated to a level high enough that the rest of its movement follows the flow of gravity. The next stop is a series of circular settling ponds, known as the "primary clarifiers," where the sludge settles out. The ponds are overtopped with skimmers to remove the oil and other floating liquids.

The sewage is then treated with a chlorine compound to disinfect it and a sodium compound to remove the chlorination. Once disinfected, the flow continues to secondary treatment, which involves aeration. At this point, bacteria take over to decompose the organic matter, converting it into a form that settles out into the middle of the aeration tank. After aeration it moves onto a third stage, where some of the organic matter is returned to the tank to provide a continuous source of bacteria.

Tertiary treatment takes place back inside a very modern building, which is entirely devoid of any sewage smells. In previous Metro facilities, before 2006, after secondary treatment, the sewage was sent to tertiary clarifiers, a final settling process, before it was discharged. Now, however, tertiary treatment involves the removal of ammonia using a sophisticated filtering process that consists of eighteen cells filled with polystyrene pellets that reduce the ammonia levels to 1 milligram per 1 liter of water. After ammonia, phosphorus is removed using a process that causes it to clump with sand. The phosphorus is then washed free from the sand, which is then recycled. The flowing water now looks completely clear because it has been treated with high doses of ultraviolet radiation, which kills any remaining pathogens. From there, the water, now drinkable, is discharged into Onondaga Lake.[4]

Another part of this process, which takes place away from the water stream, treats the biosolids that have been removed from the clarifiers. This is a two-stage process that uses chemical agents, bacteria, and centrifugal action to thicken and reduce the volume of this material by two-thirds. Water that is removed is sent back to the waste stream for treatment, and the remaining solids, or sewage sludge, is sent to a landfill.[5]

The Metropolitan Sewage Treatment plant took decades to complete, but it is an impressive operation. Its weakness involves the limits of its capacity to treat sewage. Syracuse averages over 100 inches of snow every year. When it melts, often with the help of spring rains, it can overwhelm the Metro system. When that happens, the excess sewage ends up in an overflow tank, where it is treated with chlorine and sent into the lake. If the overflow was kept in the system, it could cause violations of strict water quality standards. Ironically, in order to avoid this, the excess sewage is diverted from the three-stage process and dumped directly into the lake.

There are three ways that this could be addressed. The first would be to expand the plant's capacity to the point where it could handle the spring runoff. This would be very expensive and create an excess capacity that would be used only occasionally. The second solution would be to reconfigure the city's sewage system to separate sewage from runoff, at a cost that would be astronomical and take years of construction. The city took a third path referred to as the Save the Rain program. The program was designed to keep rainwater from flowing into the treatment plant and the surrounding creeks during heavy runoff periods, and it worked.

Save the Rain, as previously discussed, involves multiple components that engaged the entire city in the process of reducing runoff. Individuals were encouraged to acquire rain barrels and plan rain gardens. The city initiated tree plantings. Businesses were encouraged to install green roofs and cisterns. Bioswales (vegetated depressions in the landscape) were created to capture water and reduce the speed of runoff. Public and private porous paving projects were completed. This has not entirely eliminated the problem, but it has generated the additional benefits associated with greening projects, such as reducing urban heat island effects. Moreover, such projects provide a model for other urban areas.

The runoff problem isn't unique to Syracuse. Many cities that built their systems in the early part of the twentieth century combined sewer and runoff, reasoning that it was good to dilute the sewage before sending it into local rivers, streams, and lakes. The out-of-sight, out-of-mind mentality was never effective, given the nature of urban sewage, and those cities have paid a price for those decisions ever since.

A Declaration of Completion

Fish, while not edible, had for years been thriving in Onondaga Lake. In 2017, an administrator at the SUNY College of Science and Forestry (ESF) launched a project to bring Atlantic Salmon back to Onondaga Lake by stocking them in Six Mile Creek, which, through much of the twentieth century, was one of the most polluted parts of the lake's watershed. Atlantic Salmon had once been abundant in the Great Lakes region, but populations had been depleted due to overfishing, development, and contamination. The Six Mile Creek project was part of a larger effort on the part of New York State to bring back the salmon, which involved stocking them in thirteen different watersheds.[6] The initiation of the project suggested, in itself, high hopes of lake restoration.

On April 26, 2017, after decades of despoliation and years of wrangling over the outlines of a cleanup, Governor Andrew Cuomo's office disseminated a press release to announce the completion of capping on the Onondaga Lake bottom. "The completion of this project to revitalize Onondaga Lake," Cuomo stated, "marks a major milestone for Central New York that will improve the quality of life for residents, and opens up additional opportunities for tourism, outdoor recreation and economic development for the entire region. . . . This work will help ensure that the lake remains a clean, viable natural resource for generations to come." Cuomo's statement was couched entirely in the language of utilitarian calculation and economic development and implied a finality that was overstated. Still, he was right in the sense that Honeywell and the DEC had completed one of the most impressive and expensive remediation projects in American history.

The economic arguments are not without merit. The Onondaga cleanup was cast as an economic development project for politically strategic reasons. Cuomo had made economic revitalization of upstate New York one of the key elements of his time in office, an approach he referred to as "Central New York Rising." Several billion dollars in economic development funds had been provided, not only for central New York, but the southern tier, the western wing of the state, and the north country. Syracuse and other medium-sized cities that had once been among the country's most important manufacturing hubs had been decimated by deindustrialization.

The Village of Solvay was itself a casualty of shifting economic forces. The governor's emphasis on economic development was perfectly understandable, but such development, including tourism, can have negative environmental impacts. Whether this had been or would be taken into consideration going forward had not been resolved by the lake's capping. Moreover, whether Cuomo's initiative would be acceptable to the Onondaga people given their veneration of the lake was even less resolved. DEC Commissioner Basil Seggos was quoted in the press release, saying that, "In the past, Onondaga Lake was often called the most polluted lake in the nation, but today it is the most resilient."[7]

Honeywell was responsible for only part (although a very large part) of the area's contamination problems. In April 2018, the USEPA gave a Brownfield Development Grant of $200,000 to the Onondaga County Industrial Agency, to clean up the abandoned Roth Steel plant site at the southern tip of Onondaga Lake. Roth Steel, which had been in business for nearly 100 years, had been a scrap yard with a metal shredding facility. As with other industrial operations in the area, it had a troubled environmental record. In 2008, it settled with the New York Department of Environmental Conservation for improperly disposing of hazardous material. Employees at the company confirmed that in 1993, they buried 5,000 tons of such material in a swamp on the shores of Onondaga. While steel shredding can be an effective form of recycling, many non-recyclable materials remain after the shredding process is complete. If a car is shredded, what's left are the seats, the plastic, and the various chemicals necessary to make it operate. Steel recyclers need to find responsible ways to deal with this, but in Roth Steel's case, they simply dumped it into a pit. This "shredder fluff," as it is referred to in the business, can contain PCBs, cadmium, mercury, and petroleum. The company was fined $150,000 and required to clean up its mess.[8]

Meanwhile, other lakeshore development projects were going forward. In the spring of 2019, New York Lieutenant Governor Kathy Hochul and DEC Commissioner Basil Seggos broke ground for a boat launch to be constructed on Onondaga Lake, to be named after the previous DEC Commissioner Kenneth P. Lynch. In a statement, Hochul said that "The restoration of Onondaga Lake has been a remarkable success story thanks to the hard work of dedicated professionals like Kenneth Lynch, whom we proudly honor today."[9] Lynch had been DEC Commissioner for two decades, and the process of the Onondaga Lake cleanup had occurred mostly on his watch. Still, his tenure had not been without its controversies, notably in terms of his conflicts with members of the Onondaga Nation, who consistently objected to the Honeywell–DEC negotiated remediation plan.

In the summer of 2019, Onondaga County began to explore the possibility of creating a beach on Onondaga Lake and committed $330,000 to a feasibility study. Local environmentalist Lindsay Speer, a member of the Neighbors of the Onondaga Nation, commented that the beach would create an "illusion of safety." Alma Lowry, legal counsel for the Onondaga Nation, noted that the lake doesn't meet water quality standards for swimming if you take into account the multiple toxins that are buried in its sediments. She questioned how well sequestered they were and suggested that creating a beach and allowing swimming ignored potential health impacts that may be caused by exposure to them.[10] The county eventually released a proposal for a $2.8 million beach facility, which included a bathhouse, picnic areas, and a promenade. With regard to safety issues, County Environmental Commissioner Travis Glazier deferred to the findings of the regulatory authorities. "DEC, EPA, DOH have certified that Onondaga Lake is acceptable for swimming in a public bathing area, and so we're operating under that assumption," said Glazier. "People have indicated in the comment process that they don't agree with that. That's really not in the scope of what the feasibility is meant to resolve."[11]

Further initiatives for the lake were being reconsidered. In 2017, the New York DEC opened a public comment period on a series of projects, including a deep-water fishing pier in Onondaga Lake, enhancement of jetties on the lake to allow for greater access, and a boat launch on the Seneca River.[12]

Cuomo's 2017 announcement that Honeywell had finally settled with New York did not cover the company's separate arrangements with the EPA. On November 19, John McAuliffe, the project manager for the Onondaga Lake cleanup announced that, after planting about a dozen trees, Honeywell would have complied with its agreement with the State of New York. While there were still some trails and habitat restoration to complete that had been part of their agreement with the EPA, for all practical purposes, the remediation of the lake, the surrounding shorelines, and the creeks, was done. Honeywell eventually committed $6 million to settle with the EPA. In conjunction with the settlement, there was a joint press release by the US Justice Department, the New York Attorney General, and the New York DEC, that referred to the "remarkable restoration" of Onondaga Lake.[13]

Onondaga Lake and Ecological Restoration

I began this book with a description of my bicycle ride on Restoration Way, the path that now runs along the shoreline of Onondaga Lake, as I visited various contamination sites in the area. "Restoration" is a word that is often used to describe the lake's current condition, and one I have used myself in the course of writing this book. But has Onondaga Lake been "restored," and if so, in what sense? Given that Honeywell has now been deemed by the State of New York to have completed its share of the cleanup, and that it has agreed to final measures with the EPA, this seems like a good point at which to evaluate the outcomes of the process, not just in narrow technical terms like measuring the lake's methylmercury levels, but in broader historical, cultural, and ecological terms. I do not underestimate the difficulties involved with such a task, and I offer only tentative conclusions. But by examining the concept of restoration, and debates about its meaning in different contexts, I hope to offer insights useful for evaluating the status of Onondaga Lake and other ecological restoration projects.

Restoration ecologists have often drawn their conceptual maps from experiences in forested or pastoral settings. Eric Higgs begins his path-breaking book *Nature by Design* by looking at the development threats facing Jasper National Park in Alberta, Canada.[14] The restoration projects that he goes on to examine include Discovery Island in British Columbia, the Florida Everglades, and the Morava River in the Slovak Republic. Stuart K. Allison, author of the wide-ranging book *Ecological Restoration and Environmental Change*, began his career restoring prairie grass systems in the American Midwest. Their analyses of restoration processes, variously explicit and implicit, reach beyond the restoration of less intensively inhabited places, and can be drawn upon to evaluate the kinds of urban restoration projects represented by the Onondaga Lake cleanup.

Allison argues that ecological restoration involves efforts "to produce ecosystems that have health, integrity, and that are sustainable, while returning to a historical trajectory that may include dynamic changes to ecosystem structure and function, thus allowing evolutionary and ecological processes to operate."[15] In other words, ecological systems are not static. Even without human imprints, they evolve, as they have for the entirety of Earth's history. As a result, attempts to return them to a specific temporal juncture and hold them there like museum pieces are misguided. Recognizing the dynamic character of these systems, restorationists should work to return them to a condition that allows them to evolve in ways specific to their circumstances. Higgs argues that ecological restoration involves two main concepts: ecological integrity and historical fidelity. "When the complicated mix of restoration practices is sorted out, what is left is a concern for the quality of the ecosystems resulting from the restoration (integrity) and for the extent to which they reflect the history of the place (historicity)."[16]

Ecological restoration, as a practice, cannot be separated from deeper and more intractable questions involving the relationship of humans to the natural world. Long endemic to European and American cultural understandings has been a dualism that conceptually separates humans from the environments they inhabit and that implies an entitlement to dominate them. This viewpoint can be traced to a variety of historical influences, including, as cultural historian Carolyn Merchant has shown, the biblical narrative of Genesis, where humans are portrayed as managers of ecological systems, the primary purpose of which is to provide for the sustenance and prosperity of human life.[17] However, that dualistic worldview has been questioned by a diverse array of environmental thinkers, from Rachel Carson[18] to J. Baird Callicott.[19] In fact, challenging the homocentric perspective has been a central component of environmentalist thinking and the environmental movement virtually since its inception.

The philosophical and historical underpinnings of an environmental ethic can shape the practices of environmental restoration, but they don't necessarily determine them. For example, those who reject the human/nature dichotomy might be tempted toward the conclusion that humans have free rein to act solely in their own narrow interests, because, as beings within nature, they (we) are by definition acting from our naturally defined impulses. Our actions are as much a part of the natural world as those of non-human actors or forces. Pushed to its logical (or perhaps illogical) conclusion, such a viewpoint would consider a nuclear power plant (or a mercury-infused lake) and an old growth rainforest as being of equivalent ecological value, because they all reside within the natural world. That the former is designed and operated by humans makes no difference. Ultimately, in other words, there's no way to escape the natural world within which everything is embedded. If we push the logic into the arena of restoration, it becomes difficult to find a rationale for it, given the theoretical equivalence of human-organized systems and those without human intervention.

The concept of restoration is not without its critics. Robert Elliot, in his seminal essay "Faking Nature," argues that all attempts to restore ecosystems are suspect, because they apply human managerial principles to complex ecological systems that can never been fully understood, much less replicated, within the limited range of human understanding. Moreover, they provide an excuse for undermining natural systems with the false promise that they can eventually be restored.[20]

Nevertheless, restoration is an idea widely embraced by environmentalists, public officials, and citizens, not because they see themselves as separate and apart from nature, but because they feel compelled to reestablish a set of relationships with the natural world that have been broken by various forms of human intervention and economic development. A coal slag heap is not equivalent to a pristine lake, nor is a nuclear power plant equivalent to an old growth forest. The damage that humans have done to natural systems is recognized now in a way that it wasn't a hundred years ago by the humans who did the damage in the first place. This recognition and implicit guilt over our actions have spurred an interest in doing something about the messes that we have created. Partly, this is driven by human interests, economic as well as aesthetic, but I would suggest it also stems from a moral imperative that sees value in reestablishing complex ecological systems that have an integrity that exists apart from human needs and desires. Trying to disentangle these various perspectives and interests in evaluating specific restoration projects is tricky, but for efforts at restoration to be effective and have integrity, it is essential.

While all attempts at restoration are by definition at least partly managerial because they intervene or interfere in natural processes, some kinds of interference are more legitimate than others. While human interests need to be taken into account, they cannot be the sole motivation for restoration efforts. As Allison notes, "In order for restoration to avoid the problem of being just another managerial interference with nature . . . it has to avoid restoring the environment solely to benefit human needs and desires. Restoration must leave room for species that are not beneficial or desirable to humans and to allow the environment to develop along pathways that are not controlled by humans."[21] Levels of legitimate human interests will differ, given differing circumstances. In especially damaged ecosystems, where reestablishing previous pathways may be difficult to impossible, a greater emphasis on human utility maybe be allowable. Replanting native trees to reverse desertification is different than remediating a chemically contaminated waste site.

Eventually, ecosystems, no matter how damaged, will find ways to restore themselves. The mercury in Onondaga Lake will eventually be dispersed and diluted into the larger ecosystems to the point that it is undetectable, or it will be locked into the sedimentary materials and no longer carried into the lake and surrounding environs. The mounds of calcium chloride will be eroded into the landscape and carried into the lake, and at some point become a barely distinguishable aspect of the geological landscape of Central New York. But it could be decades, perhaps centuries, before the Solvay imprint is erased. The process of restoration, then, is an attempt to speed up that which would, in the long run, occur anyway. And with intervention, the final result might be closer to what would have transpired had the damage not been inflicted in the first place.[22]

While the formal term "restoration" applied to ecological settings may have a recent derivation, the practice of what we now think of restoration has a long history. Forest management or restoration projects were set up in Germany in the fourteenth century. Two thousand years ago, the Roman poet Virgil recommended that country dwellers take care of their forests. Given the disturbances that people cause, people have long been seen as an essential element in restoration or conservation efforts. In the age of the Anthropocene, the necessity of intervention takes on greater import, because little, if anything, within Earth's ecological system has not been touched by human influences in some way. Some of climate change's consequences, for example, are the greatest at the most inaccessible parts of the planet.[23]

Allison distinguishes two tracks for ecological restoration. The first is "primarily concerned with recreating previously existing historical ecosystems that had been damaged, lost, or destroyed, often working in agricultural fields or land that was set aside because it was of marginal value for development."[24] Prairie restoration projects in the American West are exemplary here, and many conservation projects that involve reintroduction of native flora and fauna could be included. The second track involves taking land that was "utterly destroyed," and which had little if any ecological function, and improving it to whatever degree possible. In the first instance, wildlife biologists might be the primary drivers for a restoration project. In the second instance, the lead may be taken by "engineers, hydrologists, and landscape architects." Here the goal is to produce "whatever kind of ecosystem was possible on that derelict land." A grassland might be turned into a marsh, a forest into an organic family farm, an industrial brownfield into an urban park.[25] The original system might not be replicated, but the restored area could still have greater ecological integrity.

Many, perhaps most, ecological restoration projects will fall somewhere in between restoring a marginal landscape and remediating an utterly destroyed local system. The Onondaga Lake project has involved elements of both. It has involved reestablishing native species of flora and fauna while also remediating highly contaminated areas along the lake's shorelines and within the lake itself. Willow trees planted in the Solvay calcium chloride beds did not replicate native species, but they were a practical way to stabilize the mounds. Bringing back Atlantic salmon and other species of fish that were eliminated when the lake was in its worst condition is an attempt to reestablish previously intact aspects of the system. Cleaning up the wastebeds and other highly contaminated areas is more like brownfield remediation.

The term restoration implies that an ecological system will be returned to some previous condition, at a time before the damage or change caused by humans was introduced. Why should this even be of concern to contemporary dwellers of a place? As Higgs argues, it suggests that the past was somehow better, representing a simpler and more coherent time. But romanticizing an imagined past can create its own set of problem, because views of the past are always interpreted through the lenses of the present. So, for example, American conservationists such as John Muir romanticized past wildernesses in North America, conveniently ignoring the social, cultural, and political contributions of the Indigenous people who inhabited it, and the violent processes by which they were removed.

Nevertheless, restoring lost ecosystems does seem to have wide appeal. As Higgs notes, "the past becomes the basis for setting forth our best judgment about the present, and in this respect will influence how in the future we will adjudicate our actions."[26] The restoration of an ecosystem has to tell a story, one persuasive enough that future generations will not only appreciate the system but protect and enhance it. This raises the crucial question of what reference point a restoration effort should attempt to replicate. In the case of Onondaga Lake, we could imagine several candidates.

One reference point would be before the development of the Solvay Process plant and the other industrial enterprises, including the metro sewage system, befouled the lake. At that time, the lake was relatively pristine. Fish caught in it could be eaten. Swimming posed no health problems. The lake's ice could be cut and sold. Yet the lake was hardly untouched. Commercial activity, in the form of the multiple resorts, existed along its shoreline. Steamboats provided transport. The lake, in other words, had the stamp of colonial occupation, and the markers of industrial development. However, it held an ecological integrity that would be extensively damaged in the twentieth century.

A second strategy would be to return the lake to its condition before white European settlers arrived, a time when the human settlement was defined by the Onondaga people; a period in which Hiawatha paddled his white canoe across Onondaga Lake and discovered the sacred wampum beads used to memorialize the Great Peace. That would involve not only remediating the polluted areas around the lake and the waters of the lake itself, but also removing the bicycling trail that now encircles it, the docks, the parks, the mall, all of the highways, and more. Such a prospect, while it may be attractive, especially to members of the Onondaga Nation, is unlikely as along as the City of Syracuse exists in its current form. Perhaps someday it won't be. Perhaps someday the land will be returned to its original inhabitants. It is impossible to predict developments that might occur decades or centuries from now. But, if that were to happen, restoration would be determined by those inhabiting the land at that time, and not by the descendants of colonial occupiers, whose practices contaminated the land in the first place.

A third, more radical form of restoration, would involve returning the area to conditions that existed before there were human settlements of any kind in North America, before the Haudenosaunee engaged in land use practices that, while more respectful of ecological sustainability, still left an imprint. To achieve such a goal is even more farfetched. For one thing, it's unclear that we have a complete enough picture of presettlement ecologies to attempt to restore them.

It is worth considering the rationale for pursuing such a radical restoration process, whatever the practical issues. It would imply that the presence of humans in an ecological system is by definition alien, unhealthy, and unwelcome. If that is the case, then it would seem that the complete removal of humans from the global ecological system, allowing it eventually to return to prehuman conditions, would be the best strategy. And the morality of doing so, given the levels of disease, starvation, and perhaps killing, that would be necessary to achieve it is, to say the least, highly questionable, to say the least. At some point, then, paths of restoration are going to have some human reference points. They must be derived from a collaboration between human and non-human natural systems, not the elimination or suppression of one or the other.

Given the ecological, political, and cultural constraints within which restoration plans must operate, some form of negotiation between human and non-human systems seems warranted. Practical considerations cannot be entirely discounted in the service of some form of ecological purity. In fact, given the complexities of ecosystems, the goal of achieving entirely pristine or untouched versions of them is probably misguided. Allison's advice along these lines seems sensible: "Ecological restoration must have some basis in the undisturbed, previous ecosystem, but we must recognize the limitations of that model for restoration and be sure that what we are attempting to do is attainable. Restoration that honors the past and allows for dynamic ecologies and evolutionary change will have the spark that defines living systems and that is the essence of 'wildness'—those aspects of nature that are beyond human control."[27] Returning some semblance of wildness to Onondaga Lake should be an important consideration, and this implies that the restoration process not be entirely driven by economic considerations or plans for future development.

While conducting my research, the phrase "urban lake" was often invoked to suggest that there were limits to how far restoration efforts could be pressed in the Onondaga case. The authors of a University of

Washington study have concluded, however, that experiencing a sense of wildness, or some degree of removal from civilization, can and should be a central feature of urban parks.[28] Having traveled on the path around Onondaga Lake, my sense is that something like this has been achieved in a small section at the northwestern end of the lake. Those areas should be protected and, if possible, expanded.

Having said that, economic considerations cannot, and probably should not, be completely discounted. As Allison notes, "Promoting economic value and social benefits of ecological restoration is especially important when restorationists work in urban settings and when they are attempting to acquire funding for restoration projects."[29] In fact, when restorations occur, they often contribute economic value to an area, and that is evident in the case of Onondaga Lake, where economic activity of various kinds, from fishing, to recreation, to entertainment, has increased as the lake cleanup has gone forward. Allison remarks, "In urban situations, ecological restoration is not likely to result in returning large areas to natural functioning, but instead will be a vital component of revitalization of urban areas."[30] In the case of Onondaga Lake, natural areas have been expanded as economic activity has increased.

While economic benefits are worth consideration, overemphasis on the economic at the expense of the ecological can undermine the purposes that inspired the restoration in the first place. The ecological protection and restoration of the lake must be kept in mind as the primary goal. Extensive commercial real estate development along the Onondaga shoreline would be inconsistent with the promises that have been made by public officials regarding the importance of restoration projects. On the positive side of the ledger, much of the land along the lakeshore is owned by Onondaga County, which provides at least some buffer (both physical and political) against extensive private development projects. Insofar as these might go forward, there would need to be some kind of negotiation with county officials with some level of public input. That input should be included multiple voices, including members of the Onondaga Nation, who have a centuries-long cultural connection to the lake and the surrounding environs.

Restoration and Indigenous Knowledge

Most restoration ecologists are in agreement on the need for balance between ecological and human interests, but how to achieve may not be obvious. The constraints to achieving balance are not simply technical or ecological, they are political. And human-organized systems of power and policy must be considered in terms of their capacities to implement meaningful and coherent restoration projects.

Restoration requires democratic participation for, as Higgs notes, "Participation is a vital evaluative component of restoration, and most restoration projects not just individuals but communities." Higgs continues, "the more participatory restoration is, the better."[31] In the case of Onondaga Lake, the decision makers with the most say were a very large global corporation and the state department of environmental conservation. It would seem only fair that the voices of the Indigenous people whose cultural traditions were most strongly tied to the lake going back for millennia, would have at least an equal voice at that table, but they didn't.

In this regard, Canada provides a potential model for how an expanded view of democratic participation could be integrated into restoration projects in the U.S., and environmental decision-making more generally. The current government in Canada has adopted a groundbreaking piece of legislation with the mundane title of the Impact Assessment Act (IAA) or Bill C-69. The bill requires consideration and incorporation of Indigenous knowledge into environmental decision-making. It will, according to the Canadian government, "introduce a requirement for the mandatory consideration of Indigenous knowledge." Canada, unlike the U.S., has historically provided an informal role for Indigenous people to have a voice in environmental policies, and Canadian courts have, in recent years, been tentatively supportive of Indigenous land claims. In 2009, the Canadian government has, for example, set up the Specific Claims Tribunal to address land claims grievances.[32] Nothing like that exists in the U.S., where courts, including in New York State, have been unresponsive, or even hostile, to historic Native American land claims and alleged treaty violations.

The IAA formalizes and expands the procedural mechanisms for having those voices heard and taken seriously, or at least that's the hope.

What constitutes Indigenous knowledge is not specified in detail in the legislation, but it considers Indigenous knowledge "in broad terms, to be inclusive of the evolving knowledge of Indigenous peoples." In other words, Indigenous knowledge is alive, vital, and grounded in the experiences, judgments, and findings of Canada's Indigenous populations. Moreover, Indigenous knowledge is used in tandem with conventional "scientific information and data."[33] (The two forms of knowledge are not incompatible, and the former can incorporate the latter.) The intent of the legislation is to expand the forms of knowledge used in environmental decision-making and recognize conflicting points of view. The IAA, along with other legislation, is designed to create what the government labels an Indigenous Policy Framework.

Given that the exact meanings of the term "Indigenous knowledge" may be open-ended, it would be self-defeating for the Canadian government to specify them in advance. Therefore, incorporating Indigenous knowledge will necessarily be a collaborative, and thus democratic, process. Given the relatively recent passage of the legislation, its impacts on environmental decision-making are yet to be determined, and will evolve, no doubt, on a case-by-case basis as Environmental Impact Statements are developed and reviewed. What is clear is that "The new Act puts a heavy emphasis on avoiding negative impacts and seeking positive impacts regarding the country's Indigenous population. This includes using what used to be called 'Traditional Knowledge,' now called 'Indigenous Knowledge.' It is based on experience and knowledge passed down through generations, but also other types of community lore as well as empirical knowledge gained through standard research methods."[34]

The Indigenous Policy Framework was part of a larger effort to remake Canada's environmental regulations and expand the scope of Environmental Impact Statements. "The focus under the IAA [Impact Assessment Act] is now broader, moving towards Impact Assessments (IAs) that consider the natural environment, but put a greater emphasis on social, health, and economic factors." In Canada, Environmental Impact Statements will have to take environmental justice into account from the start.

Given the focus of the IAA on Environmental Impact Statements, it will deal primarily with proposed projects, not the cleanup of previously damaged or contaminated sites. Canada has historically been heavily dependent on extraction industries to support its national economy: timber, fossils fuels, and minerals. Thus, much of the impetus of the Act has been to seek Indigenous participation in extraction projects. Unfortunately, implementation of the act seems to have been ignored in the case of the two British Columbia oil pipelines currently being constructed to bring crude from the Alberta Tar Sands to Pacific ports, where it can be shipped to Asia. So, while Canada may be moving in the right direction legislatively, it still has a long way to go in actual practice.[35]

In any event, similar legislation in the U.S would involve revision of the National Environmental Policy Act, and it would have broad implications for all environmental policy decisions. While I would not rule out, and would be sympathetic to such a revision, the focus in this study is narrower, dealing with U.S. and New York State Superfund laws, and incorporating Indigenous knowledge into decisions about restoring previously damaged ecosystems. But while the scope may be narrower, the rationale and same general principles apply. As the authors of one study on the potential value of Indigenous knowledge note, "IK [Indigenous Knowledge] systems offer an alternative source of knowledge, often complementary to western science. Over thousands of years of observation and culturally transmitted learning, Indigenous peoples of the world have established complex management and conservation strategies to steward local environments."[36]

What might a policy similar to Canada's IAA mean for the Onondaga Lake restoration process? The answer could be relatively straightforward. The Onondaga people, though their legal and political representatives, would have had a seat at the table when restoration plans were being developed. Their scientific studies, which questioned the data presented by the DEC and Honeywell on the extent of lake contamination and the impacts of the proposed solutions, would have had a hearing, rather than being ignored in subsequent DEC hearings and state legal rulings. Perhaps those studies would, in an open process of debate and negotiation, have been shown to be flawed. But, since they were never given an adequate hearing, it's very difficult for citizens and other interested parties to make judgments. Perhaps if the Onondaga representatives had been involved in the decision-making processes, an agreement would have been reached between all of the parties on a path forward that would have respected Onondaga cultural traditions and values. But U.S. and New York State regulations are written so as to preclude such opportunities. The result is continued marginalization of New York's Indigenous populations and a narrow definition of what restoration entails.

For evidence of the value of Indigenous participation, we need only look to Onondaga County's Save the Rain program. It had support of the Onondaga Nation and members of a variety of grassroots and institutional actors, although they did not have a formal role in proposing or implementing it. Save the Rain also had the support of Onondaga County Executive Joanie Mahoney and a favorable ruling by federal district court judge Anthony Scullin. But had the process been more open to multiple voices from the start, green solutions to the county's sewage overflow issues might have been discovered sooner, and the displacement of African American families and the disruption of their community through the siting of an industrial sewage facility in their neighborhood might have been avoided. And who knows what other past ecological violations might have been prevented or mitigated?

True restoration, in my view, would be of the kind that takes the deep cultural understandings and practices of the Onondaga Nation into account. True restoration, in other words, would need to take into account the concepts, narratives, and worldviews of the Onondaga people represented at the beginning of this book.

One final note: In June 2022, media reported that the State of New York and Honeywell agreed to return 1,000 acres of ancestral land to the Onondaga Nation. The Nation's Chief, Tadodaho Sid Hill, stated, "It is with great joy that the Onondaga Nation welcomes the return of the first substantial acreage of its ancestral homelands. The Nation can now renew its stewardship obligations to restore these lands and waters and to preserve them for the future generations yet to come." The agreement was grounded in the Onondaga Lake Natural Resource Damage Assessment and Restoration Program, an agreement between the U.S. Department of the Interior, the State of New York, and the Honeywell Corporation to provide compensation for parties injured by the despoilation of Onondaga Lake. In this case, the agreement seemed to offer a very literal form of restoration, in that it meant restoring a large tract of land to its original inhabitants. The land, which is mostly forested, includes the headwaters of Onondaga Creek, the main tributary feeding the waters of Onondaga Lake.[37]

Clan Mother Wendy Gonyea, however, was less than enthusiastic about the deal. She told me that the agreement reached by the nation's trustees was done without consultation with the larger community, and she was taken by complete surprise when the announcement was made and reported by the press. She was concerned about the stipulations that were attached to the transference.[38] The State of New York would need to sign off on any construction that was done, or other material changes to the property, and the area would have to be open to public access.

It seems that the state was giving title to the land to the Onondagas, but it was not offering them actual control or true ownership of it. It's difficult to avoid the conclusion that a colonial mentality still pervaded state and corporate officials involved in the negotiations. Given objections raised by members of the Onondaga community, after the announcement was made, the status of the deal was put into limbo. And, given the Onondaga's consensus-based decision-making process, it would take time to resolve differences. At that point, negotiations with the state might need to be reopened. It's unclear how the state might react.

As with other questions related to the Onondaga Lake cleanup, here things turned out to be much more complicated than they at first appeared.

ACKNOWLEDGMENTS

I spent longer on this project than any of my previous research efforts. This was partly due to other projects and events intervening while I was at work on it, but also due to the wealth of historical and contemporary materials that were available. Since much of the research involved complex scientific findings, I needed to spend considerable time familiarizing myself with some of the complexities of industrial chemistry and toxicology. When a project extends for such a long period of time, there are numerous people who provide support and advice along the way. I wanted to be sure to recognize them for it.

Ithaca College provided two small grants at the start of the project that helped me initiate the research process. The Ithaca College Interlibrary Loan staff has been extremely helpful throughout, tracing down sometimes obscure materials and providing them to me in timely matter. Sarah Shank deserves special mention here. The librarians at the Solvay Public Library, housed in a beautiful Carnegie building in Solvay, New York, provided me with guidance and access to their excellent archival collection, which included materials on the original Solvay plant siting and the broader history of Allied Chemical. They have also collected technical documents related to the cleanup.

My friend and colleague Richard Buttny, Professor Emeritus in the Department of Communication and Rhetorical Studies at Syracuse University, read parts of the manuscript, provided useful criticisms, aided me in numerous conversations over dinner, and introduced me to others in the area who were also helpful.

I much appreciated the generosity of my interviewees. They provided diverse, sometimes conflicting, points of view, and contributed insights that I would have been unable to uncover from written sources alone.

David Matthews, Director of the Upstate Freshwater Institute, successor to legendary local environmentalist Steve Effler, walked me through technical and historical aspects of the cleanup effort. Charles Driscoll,

University Professor of Environmental Systems and Distinguished Professor of Civil and Environmental Engineering at Syracuse University, and a world renowned expert on mercury contamination, patiently explained to me the complexities of methylmercury chemistry and his views on the successes of the cleanup effort.

Tim Minnich, partner in the firm, Minnich and Scotto, provided detailed information on technical and legal matters related to the Camillus waste site. Lynda K. Wade, Camillus resident and a founder of the Camillus Clean Air Coalition, helped me understand impact of the dredging operation on her community and provided extensive documentation to support her claims.

Civil rights attorney Joseph Heath, General Counsel for the Onondaga Nation, provided an important technical and legal critique of the cleanup effort and encouraged me to question the official story on the effort's success. Onondaga Nation Clan Mother Wendy Gonyea generously invited me to the Onondaga Territory for a lengthy conversation on the history and culture of the Onondaga people and explained the significance of (what are now) Central New York's watersheds and broader ecological systems to their community. Tehosterihens Deer, a Canadian student of Richard Buttny's and a member of the Mohawk First Nations community, encouraged me to consider possible alternatives to current U.S. approaches to inclusion (i.e., exclusion) of Indigenous perspectives in environmental policy decisions.

Joanie Mahoney, currently President of the SUNY College of Environmental Sciences and Forestry, and former Onondaga County Executive, took time to explain the Save the Rain program, for which she had been a strong advocate when serving in county government. The staff at the Onondaga Metropolitan Treatment Plant provided an excellent and fascinating tour of their now state-of-the-art facility, which involved a walk-through, and detailed explanation of every phase of the operation.

This project, as is the case with all my efforts, was enhanced in innumerable, if not always direct, ways by my long-running conversations with my friends and colleagues at Ithaca College, including Professors Don Beachler, Zillah Eisenstein, Gordon Rowland, Anne Stork, and Patricia Zimmermann.

Jane Banks, a "recovering academic" and freelance editor extraordinaire, did a first-rate job of editing a draft of the manuscript, providing substantive feedback and encouraging me to improve the organization and clarity of the presentation.

NYU Press Executive Editor Ilene Kalish shepherded the manuscript through the review process and offered feedback that led to significant improvements in the final draft. Editorial Assistant Yasemin Torfilli guided me, and the completed manuscript, through the production process.

NOTES

INTRODUCTION

1 It's not entirely clear how, or from whom, the phrase originated, or whether it was true, but the phrase stuck as an unfortunate tagline to Onongaga Lake. David Chanatry, "America's 'Most Polluted' Lake Finally Comes Clean," NPR: All Things Considered, July 31, 2012, www.npr.org.

2 For purposes of narrative simplicity, I combined several trips to the area into the one recounted here.

3 Donald L. Pair, "Geomorphic Studies of Landslides in the Tully Valley, New York: Implications for Public Policy and Planning," *Geomorphology* 47, no. 2 (2002): 125–135.

4 "Onondaga Creek Mudboil Study," Onondaga Nation: People of the Hills website, www.onondaganation.org; D. L. Pair, W. M. Kappel, and M. S. Walker, "History of Landslides at the Base of Bare Mountain, Tully Valley, Onondaga County, New York," United States Geological Survey Fact Sheet 0190-99, June 2000; United States Geological Survey, "Remediation of Mudboil Discharges in the Tully Valley in Central New York," 2009, https://pubs.er.usgs.gov. Accessed August 4, 2010.

5 Indigenous Values Initiative, "Wampum," https://indigenousvalues.org. Accessed January 26, 2023.

6 Wendy Gonyea, interviewed by author, March 15, 2022.

7 "Long Branch Park at Onondaga Lake Park," Onondaga County Parks website, www.onondagacountyparks.com. Accessed August 5, 2020.

8 Gonyea interview.

9 Minnich and Scotto, Inc., "Response to DEC Review of Minnich and Scotto's April 8, 2013 Report: 'Air Contaminant Exposure to Residents of the Town of Camillus from Honeywell's Sediment Treatment and Containment Facility,'" May 8, 2013. Email attachment to author, August 3, 2021.

10 J. Donald Hughes, *An Environmental History of the World: Humankind's Changing Role in the Community of Life* (New York: Routledge, 2001), 4.

11 Andrew J. Isenberg, "Introduction: A New Environmental History," *The Oxford Handbook of Environmental History*, ed. Andrew Isenberg (Oxford: Oxford University Press, 2014), 1.

12 J. R. McNeill, *Plagues and Peoples* (New York: Anchor Books, 1976).

13 J. R. McNeill and Peter Engelke, *The Great Acceleration: An Environmental History of the Anthropocene since 1945* (Cambridge, MA: The Belknap Press, 2014), 211.

14 Carolyn Merchant, *Ecological Revolutions: Nature, Gender, and Science in New England* (Chapel Hill: University of North Carolina Press, 2010).

15 Carolyn Merchant, *The Death of Nature: Women, Ecology, and the Scientific Revolution* (New York: HarperOne, 2019).

16 Carolyn Merchant, *Reinventing Eden: The Fate of Nature in Western Culture* (New York: Routledge, 2013).

17 Carolyn Merchant, *Earthcare: Women and the Environment* (New York: Routledge, 1995).

18 Carolyn Merchant, *The Anthropocene and the Humanities: From Climate Change to a New Age of Sustainability* (New Haven, CT: Yale University Press, 2020).

19 Stephen Mosley, *The Environment in World History* (New York: Routledge, 2010), 3.

20 David Stradling, *The Nature of New York: An Environmental History of New York State* (Ithaca, NY: Cornell University Press, 2010), 1; for references to Onondaga Lake, see 177, 203, 217, 229.

21 Larry G. Johnson, *Tar Creek: A History of the Quapaw Indians, the World's Largest Lead and Zinc Discovery, and the Tar Creek Superfund Site* (Norman, OK: Tate Publishing 2008).

22 Joseph Alexiou, *Gowanus: Brooklyn's Curious Canal* (New York: New York University Press, 2015).

23 Ibid., 340.

24 Anthony N. Penna and Conrad Edick White, eds., *Remaking Boston: An Environmental History of the City and Its Surroundings* (Pittsburgh: University of Pittsburgh Press, 2009).

25 Philip Shabecoff, *Earth Rising: American Environmentalism in the 21ˢᵗ Century* (Washington, DC: Island Press, 2000), 17.

26 Rachel Carson, *Silent Spring* (New York: Houghton Mifflin, 2003).

27 See Elizabeth D. Blum, *Love Canal Revisited: Race, Class, and Gender in Environmental Activism* (Lawrence: University of Kansas Press, 2008); Lois Marie Gibbs, *Love Canal: And the Birth of the Environmental Health Movement* (Washington, DC: Island Press, 2011); Richard S. Newman, *Love Canal: A Toxic History from Colonial Times to the Present* (Oxford: Oxford University Press, 2019).

28 Eileen McGurty, *Transforming Environmentalism: Warren County, PCBs, and the Origins of Environmental Justice* (New Brunswick, NJ: Rutgers University Press, 2006).

29 Robert D. Bullard, *Dumping in Dixie: Race, Class, and Environmental Quality* (New York: Routledge, 2000). See also Robert D. Bullard, ed., with foreword by Benjamin Chavis, Jr., *Confronting Environmental Racism: Voices from the Grassroots* (Brooklyn: South End Press, 1999).

30 Merchant, *Reinventing Eden*.

31 Kimberly K. Smith, *African American Environmental Thought: Foundations* (Lawrence: University of Kansas Press, 2007).

32 Merchant, *Earthcare*.

33 Linda Lear, "Introduction" to Carson, *Silent Spring*, x.

34 Lorraine Boissoneault, "The Deadly Donora Smog of 1948 Spurred Environmental Protection—But Have We Forgotten the Lesson?" *Smithsonian Magazine*, October 26, 2018, www.smithsonianmag.com; Devra Lee Davis, *When the Smoke Ran Like Water: Tales of Environmental Deception and the Battle Against Pollution* (New York: Basic Books, 2003).

35 Ellen Griffith Spears, *Rethinking the Environmental Movement Post-1945* (New York: Routledge, 2020), 79.

36 Chad Montrie, *The Myth of* Silent Spring: *Rethinking the Origins of American Environmentalism* (Berkeley: University of California Press, 2018), 16–17.

37 Ibid., 56.

38 Kenneth A. Gould et al., *Local Environmental Struggles: Citizen Activism in the Treadmill of Production* (Cambridge: Cambridge University Press, 1996).

39 Carol Hager and Mary Alice Haddad, eds., *NIMBY Is Beautiful: Cases of Local Activism and Environmental Innovation around the World* (New York: Berghahn Books, 2015).

40 Nikolay Mihaylov and Douglas Perkins, "Local Environmental Grassroots Activism: Contributions from Environmental Psychology, Sociology and Politics," *Behavioral Sciences* 5, no. 1 (2015): 121–153.

41 Robert Vickers, "Thinking Globally Acting Locally: An Overview of Local Environmental Activism in Britain" (Ph.D. Thesis, Loughborough University, 2013).

42 Allen Hassaniyan, "The Environmentalism of the Subalterns: A Case Study of Environmental Activism in Eastern Kurdistan/Rojhelat," *Local Environment* 26, no. 8 (2021): 930–947.

43 Emmanuel Yamoah Tenkorang, "Interactions between Traditional and Modern Institutions in Environmental Governance of Mining in Asutifi North District of Ghana," *Local Environment* 26, no. 11 (2021): 1313–1329; Hyerim Yoon, Elena Domene, and David Sauri, "Assessing Affordability as Water Poverty in Metropolitan Barcelona," *Local Environment* 26, no. 11 (2021): 1330–1345; Noriko Otsuka, Hirokazu Abe, Yuto Isehara, and Tomoko Miyagawa, "The Potential Use of Green Infrastructure in the Regeneration of Brownfield Sites: Three Case Studies from Japan's Osaka Bay Area," *Local Environment* 26, no. 11 (2021): 1346–1363; Md Nurul Amin, Md Asaduzzaman, Alamgir Kabir, Shaila Sharmin Snigdha, and Md Sarwar Hossain, "Lessons from Local Indigenous Climate Adaptation Practices: Perceptions and Evidence from Coastal Bangladesh," *Local Environment* 26, no. 8 (2021): 967–984.

44 See Paul Sabin, *Public Citizens: The Attack on Big Government and the Remaking of American Liberalism* (New York: W. W. Norton, 2021).

1. ORIGINS

1 J.N.B. Hewitt, "An Onondaga Version," *Iroquoian Cosmology* (London: Forgotten Books, 2008 [1903]), 13–15.

2 Ibid., 15–17.

3 Ibid., 22–23.
4 Ibid., 23–25.
5 Ibid., 24–25.
6 Ibid., 32–33.
7 Ibid., 33–34.
8 Ibid., 37
9 Ibid., 38.
10 Ibid., 39–40.
11 Ibid., 43.
12 Ibid., 43–48.
13 Ibid., 48.
14 Bruce Elliot Johnson and Barbara Alice Mann, eds., *Encyclopedia of the Haudeno-saunee (Iroquois Confederacy)* (Westport, CT: Greenwood Press, 2000), 85.
15 William N. Fenton, "This Island, the World on the Turtle's Back," *Journal of American Folklore* 75, no. 298 (1962): 298.
16 Lee Irwin, *Coming Down from Above: Prophecy, Resistance, and Renewal in Native American Religions* (Norman: University of Oklahoma Press, 2008), 148.
17 Henry Wadsworth Longfellow, *The Song of Hiawatha* (Chicago: J. G. Ferguson Co., 1968), 4–5.
18 Ibid., 216.
19 Ibid., 226.
20 Elias Lönnrot, *Kalevela: An Epic Poem after Oral Tradition* (Oxford: Oxford University Press, 2009).
21 "Introductory Note," *The Song of Hiawatha*, iv.
22 Ibid., iv–v.
23 *Encyclopedia of the Haudenosaunee*, s.v. "Hiawatha (Ayonwantha) (Mohawk)."
24 Ibid.
25 Alan Trachtenberg, *Shades of Hiawatha: Staging Indians, Making Americans, 1880–1930* (New York: Hill and Wang, 2004), 32–33.
26 Posthumously, Wordsworth received a writing credit for three films: *Little Hiawatha*, a Disney short made in 1937; *Hiawatha*, made in 1952 with Vince Edwards; and *The Song of Hiawatha*, made in 1997 starring Litefoot and Adam Beach (both Indigenous).
27 Quoted in Mark Shepard's "Introduction to Horatio Hale: Hiawatha and the Iroquois Confederation," www.markshep.com.
28 Horatio Hale, *Hiawatha and the Iroquois Confederation: A Study in Anthropology* (Salem, MA: Salem Press, 1881), www.gutenberg.org.
29 Ibid.
30 Ibid.
31 Thomas R. Henry, *Wilderness Messiah: The Story of Hiawatha and the Iroquois* (New York: Bonanza Books, 1955), 35.
32 Ibid., 36–37.

33 Wendy Gonyea, interviewed by author, March 11, 2022.

34 Henry, *Wilderness Messiah*, 40.

35 Ibid., 42–43.

36 Ibid., 43.

37 *Encyclopedia of the Haudenosaunee*, s.v. "Hewitt, John Napoleon Brinton (Tuscarora)."

38 Ibid., s.v. "The Peacemaker (Deganawida).

39 Ibid., s.v. "The Second Epoch of Time."

40 Henry, *Wilderness Messiah*, 275–276.

41 Ibid., 277–278.

42 Gonyea interview.

43 Henry, *Wilderness Messiah*, 282.

44 Barbara Mann and Jerry L. Fields, "A Sign in the Sky: Dating the League of the Haudenosaunee," *American Indian Culture and Research Journal* 21, no. 2 (1997): 105–163.

45 *Encyclopedia*, s.v., "Government Functioning and Powers of the Haudenosaunee League."

46 Ibid., s.v. "The Second Epoch of Time."

47 "The confederation of the original 13 colonies into one republic was influenced by the political system developed by the Iroquois Confederacy, as were many of the democratic principles which were incorporated into the constitution itself." U.S. Congress, Senate, Select Committee on Indian Affairs, 100th Cong. 2d sess., 1988, H. Con. Res. 331, 1–2, www.senate.gov.

48 Carolyn Merchant, *Reinventing Eden: The Fate of Nature in Western Culture* (New York: Routledge, 2013).

49 Gary D. Rosenberg, "Introduction: The Revolution in Geology from the Renaissance to the Enlightenment," in *The Revolution in Geology from the Renaissance to the Enlightenment*, ed. Gary D. Rosenberg (Boulder, CO: The Geological Society of America, 2009), 1–11.

50 Thomas Kuhn, *The Structure of Scientific Revolutions* (Chicago: University of Chicago Press, 2012).

51 Gary D. Rosenberg, "The Measure of Man and Landscape in the Renaissance and Scientific Revolution," in *The Revolution in Geology from the Renaissance to the Enlightenment*, ed., Gary D. Rosenberg (Boulder, CO: The Geological Society of America, 2009), 14.

52 Ibid., 14.

53 Ibid., 14–15.

54 Kuang-Tai Hsu, "The Path to Steno's Synthesis on the Animal Origin of Glossopetrae," in *The Revolution in Geology*, ed. Rosenberg, 93–106.

55 Thomas Clark and Colin Stearn, *The Geology of North America: A Regional Approach to Historical Geology* (New York: The Ronald Press Co., 1969), 11.

56 Ibid., 12.

57 See, for example, Robert A. Stafford, *Scientist of Empire: Sir Roderick Murchison, Scientific Exploration and Victorian Imperialism* (Cambridge: Cambridge University Press, 2002).

58 John Wilson and Ron Clowes, *Ghost Mountains and Vanished Oceans: North America from Birth to Middle Age* (Toronto: Key Porter Books, 2009), 178.

59 Christopher R. Scotese, "Paleomap Project," www.scotese.com. Accessed August 16, 2010.

60 Steven M. Stanley, *Earth System History* (New York: W. H. Freeman and Company, 2005), 258–260.

61 O. D. von Engeln, *The Finger Lakes Region: Its Origin and Nature* (Ithaca, NY: Cornell University Press, 1962), 2.

62 David C. Roberts, *Geology: Eastern North America* (Boston: Houghton Mifflin Co., 1996), 58.

63 Ibid., 120–124,

64 Yngvar W. Isachsen, *Geology of New York: A Simplified Account* (Albany: New York State Museum, 2000), 113–114.

65 Philip F. Schneider, *Notes on the Geology of Onondaga County, N. Y.* (Syracuse, Hall & McChesney, 1894), 16.

66 Ibid., 28.

67 Isachsen, 116.

68 Schneider, 26.

69 Ibid., 36.

70 Isachsen, 163–171.

71 U.S. Geological Survey, "Paleoenvironmental Assessment and Deglacial Chronology of the Onondaga Trough, Onondaga County, New York," U.S. Geological Survey, U.S. Department of the Interior, June 2007. https://pubs.usgs.gov. Accessed May 2, 2017.

72 "Salt in Syracuse that Dug the Canal," Young American Republic website, http://projects.leadr.msu.edu. Accessed May 14, 2020.

2. TRANSFORMING A LANDSCAPE

1 Maxime Rapaille, *Solvay: A Giant*, trans. Romilly Harisson (Brussels: Didier Hatier, 1990), 29.

2 David M. Kiefer, "Chemistry Chronicles: It Was All about Alkali," *Today's Chemist at Work*, 2002, https://pubs.acs.org. Accessed June 17, 2017.

3 Rapaille, 33.

4 "The Leblanc Process," Wikipedia, https://en.wikipedia.org. Accessed June 15, 2017.

5 Rapaille, 31.

6 "Nicholas Leblanc: French Chemist and Surgeon," *Chemistry Explained*, www.chemistryexplained.com. Accessed February 20, 2017.

7 Kenneth Bertrams, Nicolas Coupain, and Ernst Homburg, *Solvay: History of a Multinational Family Firm* (Cambridge: Cambridge University Press, 2013), 12–13.

8   See P.J.A. Shaw and W. Halton, "Classic Sites: Nob End, Bolton," *British Wildlife* 10 (1998): 13–17.

9   Bertrams, Coupain, and Homburg, 13.

10 "Fresnel Lantern," Wikipedia, https://en.wikipedia.org. Accessed December 2, 2021.

11 Bertrams, Coupain, and Homburg, 14–15.

12 Malcolm W. H. Peebles, *Evolution of the Gas Industry* (London: Palgrave, 2014).

13 Bertrams, Coupain, and Homburg, 15–17.

14 Rapaille, 27.

15 Ibid., 28.

16 Bertrams, Coupain, and Homburg, 21–27.

17 Ibid., 28–29.

18 Ibid., 36–42.

19 "Solvay Process," *Encyclopedia of Science*, www.daviddarling.info. Accessed February 27, 2017.

20 Ibid., 41.

21 Ibid., 43–44.

22 William B. Cogswell, *Mining and Metalurgy* 176 (August 1921): 25–26.

23 Edward N. Trump, *Early History of the Solvay Process, Vol. 1*, 3, Solvay Public Library, https://solvaylibrary.org.

24 Ibid., 3.

25 Bertrams, Coupain, and Homburg, 48.

26 Trump, 3.

27 Ibid., 6.

28 Kathy Crowell, "History of the Town of Geddes," in Dwight H. Bruce, ed., *On-ondaga's Centennial*, Vol. 2 (Boston: Boston History Company, 1896), 1037–1048. www.rootsweb.ancestry.com. Accessed March 9, 2017. Trump, 10.

29 Trump, 8.

30 Ibid., 25.

31 Ibid., 10, 11.

32 Ibid., 11–14, 20, 21–22.

33 Ibid., 16.

34 Ibid., 31.

35 Ibid., 31.

36 Ibid., 39.

37 Fred Aftalion, *A History of the International Chemical Industry: From the "Early Days" to 2000* (Philadelphia: Chemical Heritage Press, 2001).

38 Cogswell, 26.

39 Solvay's biographer, Maxime Rapaille, seems to suggest that Semet-Solvay was founded to produce coke *before* Solvay Process was established in the U.S. (Rapaille, 23). This seems to be a mistake, given that Semet-Solvay wasn't founded until 1895, several years after Solvay Process had been established. But the book is hardly a work of analytical history, and, given its laudatory character, it is more like a long press release.

40 John Fulton, *Coke: A Treatise on the Manufacture of Coke and the Saving of By-Products* (Scranton, PA: The Colliery Engineer Co., 1905), 133–134.

41 Ibid., 134–135.

42 Ibid., 136.

43 Maurice J. Monti, *History of the Honeywell Corporation*, www.hon-area.org. Accessed March 27, 2017.

44 Allen W. Hatheway, *Remediation of Former Manufactured Gas Plants and Other Coal-Tar Sites* (Boca Raton, FL: CRC Press, 2011), 203–204.

45 Y Hu et al., "Increased Risk of Chronic Obstructive Pulmonary Diseases in Coke Oven Workers: Interaction between Occupational Exposure and Smoking," *Thorax* 61, no. 4 (2006): 290–295.

46 C. K. Redmond, "Cancer Mortality among Coke Oven Workers," *Environmental Health Perspectives* 52 (1983): 67–73, https://ehp.niehs.nih.gov; U.S. E.P.A., *An Assessment of the Health Impacts of Coke Oven Emissions*, April 1978, https://nepis.epa.gov.

47 Bertrams, Coupain, and Homburg, 48–49.

48 Richard Holmes, *The Age of Wonder: The Romantic Generation and the Discovery of the Beauty and Terror of Science* (New York: Vintage, 2010), 235–304, 337–381.

49 Alan W. Hirshfeld, *The Electric Life of Michael Faraday* (London: Walker Books, 2006).

50 David M. Kiefer, "When Industry Charged Ahead," *Today's Chemist at Work*, https://pubs.acs.org. Accessed March 2, 2017.

51 "Werner von Siemens," Wikipedia, https://en.wikipedia.org. Accessed March 2, 2017.

52 "Zénobe Gramme," Wikipedia, https://en.wikipedia.org. Accessed March 2, 2017.

53 Kiefer.

54 "Castner-Kellner Process," Wikipedia, https://en.wikipedia.org. Accessed March 1, 2017; Bertrams, Coupain, and Homburg, 123–127.

55 Bertrams, Coupain, and Homburg, 134–135, 137.

56 See Franklin Lambert and Frits Berends, *Einstein's Witches' Sabbath and the Early Solvay Councils: The Untold Story* (Les Ulis, France: EDP Sciences, 2022).

57 Ibid., 140.

58 Ibid., 99.

59 Rapaille, 11.

60 Bertrams, Couplain, and Homburg, 99.

61 Rapaille, 15.

62 Bertrams, Couplain, and Homburg, 99.

63 "Hazard Family," Wikipedia, https://en.wikipedia.org. Accessed March 8, 2017.

64 Rita Cominolli, *Smokestacks Allegro: The Story of Solvay, A Remarkable Industrial/Immigrant Village (1880–1920)* (New York: Center for Migration Studies, 1990).

65 Ibid., 105–106.

66 Donald H. Thompson, *The Golden Age of Onondaga Lake Resorts* (Fleischmanns, NY: Purple Mountain Press, 2002), 22–32.

67 Thompson, 33, 42.
68 Ibid., 43–51.
69 Ibid., 60–70.
70 Ibid., 71–80.
71 Ibid., 81–89.
72 Ibid., 105–126.
73 Ibid., 52–59.
74 Cominolli, 106
75 Ibid., 49.
76 Ibid., 49.
77 Ibid., 106–110.
78 Ibid., 49.
79 Ibid., 100–101.
80 Judith LaManna Rivette, *Solvay Stories: A One Hundred Year Diary of Solvay, New York, Its Days and Its People* (Liverpool, NY: Oh, How Upstate, 2003).
81 Judith LaManna Rivette, *Solvay Stories II: More from the Diary of Solvay, New York* (Liverpool, NY: Oh, How Upstate, 2004).
82 Cominolli, 138.
83 Ibid., 138–140.
84 Ibid., 140.

3. POLLUTE THE LAKE, SAVE THE LAKE

1 Roy Fairman, "Mayor Baldwin Saw Bright Future for Syracuse in 1846 Talk," *Syracuse Herald-Journal*, July 20, 1951, 2.
2 Grace Lewis, "Onondaga Lake Bothers Conscience of Syracuse," *Post-Standard*, September 13, 1953, A4.
3 "Survey Shows Wastebed Leakage," *Post-Standard*, June 30, 1945, 11.
4 "Stop Pollution," *Syracuse Herald*, November 8, 1911, 4.
5 "Solvay Firm Gets Blasting at Liverpool," *Syracuse Herald-American*, September 16, 1945, 26.
6 Karl Imhoff, *Handbook of Urban Drainage* (Hoboken, NJ: Wiley Interscience, 1989).
7 Glenn D. Holmes and W. P. Gyatt, "Syracuse Sewage Treatment Works: Summary of Four Years Operating Experience," *Sewage Works Journal* 1 (1929): 318–322.
8 Holmes and Gyatt, 328.
9 Ibid., 325–327.
10 "Property Owner Sues Solvay Company for Lake Pollution," *Syracuse Herald*, May 22, 1928, 3.
11 "Vision Great Parkway within Five Years," *Syracuse Herald*, May 13, 1928, 2.
12 Lewis, A4.
13 "Cleaning up Onondaga Lake Sought," *Syracuse Herald Journal*, March 8, 1942, 25.
14 "Solvay Firm Gets Blasting at Liverpool," *Syracuse Herald-American*, September 16, 1945, 26.

15 "Survey Shows Wastebed Leakage," *Post-Standard*, June 30, 1946, 11.

16 "Lake Clean-up to Be Pushed," *Post-Standard*, June 16, 1946, 17.

17 "Onondaga Lake Pollution Check, June 16," *Syracuse Herald-Journal*, May 20, 1946, 8.

18 "Citizens Push Lake Waste Fight: Solvay Co. Gets Time for Reply," Syracuse Herald- Journal, June 16, 1946, 1.

19 "Solvay Process," *Post-Standard*, June 23, 1946, 17, 22.

20 "Plenty of Carp Found by Biologists at Lake," *Post-Standard*, June 20, 1946, 17.

21 "State Probe of Onondaga Lake Proposed," *Post Standard*, June 18, 1946, 20.

22 "Restore Lake Beauty," *Syracuse Herald-Journal*, May 22, 1946, 18.

23 "Tour Lake on Pollution Inspection," *Syracuse Herald-Journal*, July 19, 1946, 19.

24 "Brand Solvay Lake Filth 'Imposition,'" *Syracuse Herald-Journal*, July 20, 1946, 3.

25 "Dead Clause Cited in Lake Land Case," *Syracuse Herald-American*, August 4, 1946, 34.

26 "Officials Survey Lake for Pollution," *Syracuse Herald-Journal*, August 16, 1946. 15.

27 "Lake Must Be Restored," *Syracuse Herald-Journal*, September 18, 1946, 20.

28 "Heated Charges Fly at Pollution Hearing," *Syracuse Herald-Journal*, September 19, 1946, 22.

29 "Keep Present Site, Welch Urges," *Syracuse Herald-American*, September 8, 1946, 38.

30 "Fair Grounds as Factory Site Urged," *Syracuse Herald-American*, February 6, 1947, 48.

31 "Must Prevent Pollution," *Syracuse Herald-American*, September 27, 1947, 35.

32 "Tell the Facts," *Syracuse Herald-American*, October 31, 1947, 32.

33 "Blames Bad Judgement for Pollution," *Syracuse Herald-Journal*, October 30, 1947, 22.

34 "Deadline Set for Action on Pollution," *Syracuse Herald-American*, November 9, 1947, 38; "Onondaga Lake Group Demands State Enforce Health Laws," *Post-Standard*, November 9, 1947, 22.

35 "Statutes to Be Invoked if Parleys Fail to Bring Lake Clean-up," *Post-Standard*, February 15, 1948, 20.

36 "Lake Pollution Must End," *Herald American*, July 4, 1948, 25.

37 Grace R. Lewis, "Authors of Pollution Shirk Guilt: No Plan Drafted to Halt Dumping Waste into Lake," *Post-Standard*, July 18, 1948, 19.

38 "Puts Pollution up to Dewey," *Post Standard*, September 17, 1948, 38.

39 "Solvay Dust, Fumes, Increase Is Charged," *Syracuse Herald-Journal*, September 30, 1948, 8.

40 "Air Pollution Hearing Held," *Syracuse Herald-Journal*, October 7, 1948, 4.

41 "Notices Out Today to All Violators of Smoke Ordinance," *Post-Standard*, September 16, 1948, 22.

42 "Solvay People Entitled to Clean Air," *Syracuse Herald-Journal*, October 1, 1948, 28.

43 "Other Solvay Industries Blamed for Air Pollution," *Post-Standard*, October 7, 1948, 8.

44 "New Solvay Equipment to Halt Escaping Fly-Ash, Lessen Lime Dust," *Syracuse Herald-American*, November 7, 1948, 1C.

45 "Ending Air Pollution," *Syracuse Herald-American*, November 5, 1948, 30; "Success in Sight," *Syracuse Herald-American*, December 19, 1948, 46.

46 "Control Commission on Pollution Urged," *Syracuse Herald-American*, December 19, 1948, 1C.

47 "Solvay Process Will Halt Overflow into Lake," *Post-Standard*, February 2, 1949, 1.

48 "Solvay," *Syracuse Herald-Journal*, February 2, 1949, 8.

49 "What Future Has Onondaga Lake?" *Post-Standard*, February 3, 1949, 30; "Cooperation Vital in Pollution War," *Syracuse Herald-Journal*, February 3, 1949, 18.

50 "Waste Plan May Mean New City Facilities," *Syracuse Herald-Journal*, February 2, 1949, 8.

51 "Solvay Process Shift Held Possible," *Post-Standard*, June 12, 1949, 1, sec. III.

52 "Welch Hurls Challenge at West Side Group," *Post-Standard*, June 27, 1949, 9.

53 "There's Another Side to the Question," *Post-Standard*, June 29, 1949, 25.

54 "West Side Association Criticizes Welch for His 'Confusing' Attack," *Post-Standard*, July 3, 1949, 16.

55 "Maloney Hits Employment Defense in Lake Pollution; See Reclamation," *Post-Standard*, July 3, 1949, 16.

56 Walter Welch, "Let Them See for Themselves," *Post-Standard*, July 10, 1949, 22.

57 "They Want Work, Not the Lake," *Post-Standard*, July 24, 1949, 20.

58 Solvay Process Advertisement, *Post-Standard*, October 9, 1949, 2.

59 Walter Welch, "They Are Silent on These Issues," *Syracuse Post-Standard*, October 16, 1949, 23.

60 "Solvay," *Syracuse Herald-American*, November 20, 1949, 37.

61 "Deeds 2 Minutes Apart," *Post-Standard*, November 20, 1949, 25.

62 "Board Votes," *Syracuse Herald-Journal*, December 1, 1949, 1, 4.

63 "Lake Clean-up Discussed by Experts," *Syracuse Herald-American*, December 8, 1949, 22.

64 "Solvay Plan Rejection Is Commended," *Syracuse Herald-Journal*, January 3, 1950, 3.

65 "Citizens Unit to Report on Wastebed," *Post-Standard*, January 17, 1950, 11.

66 "Solvay," *Syracuse Herald-Journal*, January 17, 1950, 8.

67 "Open Meeting Slated on Wastebeds," January 19, 1950, 12.

68 "Maloney Replied in Solvay Dispute," *Syracuse Herald-Journal*, January 28, 1950, 20.

69 "Officials of City to Seek Sewage Crisis Solution," *Post-Standard*, June 18, 1950, 33.

70 "Chlorine for Sewage Freed by Strikers," *Syracuse Herald-Journal*, June 21, 1950.

71 "$95 is Daily Cost to Purify City Sewage," *Syracuse Herald-American*, 18, June 1950, 40; "Wastebed Dumping of Sewage Ok'd," *Syracuse Herald-Journal*, June 15, 1950, 2.

72 "Liverpool P.T.A. Asks Quick Action on Lake Pollution," *Post-Standard*, June 30, 1950, 19.

73 "West Side Sewage System Planned," *Post-Standard*, July 2, 1950, 1, II.

74 "State Board Gives Views on Pollution," *Syracuse Herald-Journal*, June 29, 1950, 14; "Corcoran 'Crew' Make Lake Study," *Syracuse Herald-Journal*, June 30, 1950, 22.

75 "Carp Alone Can Survive Pollution," *Syracuse Herald-Journal*, June 26, 1950, 24

76 "City Officials View Sewage Problem," *Post-Standard*, September 20, 1950, 8.

77 "City Resumes Dumping at Lakeland," *Syracuse Herald-Journal*, October 26, 1950, 1.

78 "Pope Protesting State's Order on Sewage Disposal," *Post-Standard*, October 17, 1950, 4.

79 "State Okays Sewage Sludge Beds," *Post-Standard*, October 26, 1950; "Geddes Forbids Dumping by City, Solvay Process," *Post-Standard*, October 31, 1950, 6; "Pope Plans Action to Halt Sewage Dumping," *Post-Standard*, November 1, 1950, 3.

80 "Geddes Worsens Sewage Problem," *Post-Standard*, November 1, 1950, 22.

81 "Resolution States Plan to Hold Vote on Sewage Plant," October 31, 1950, 6.

82 Nelson F. Pitts, "Onondaga Lake Today," *Herald-American*, July 29, 1951, 23.

83 William A. Mahoney, "Pure Water Issue for Onondaga Lake First Essential," *Post-Standard*, December 23, 1951, 9.

84 "Pitts Cites Need for New Sewage Plant," *Post-Standard*, January 11, 1952, 14.

85 "Conference Today May Bring Action on Sewage Setup," *Post-Standard*, January 25, 1952, 36.

86 "Officials to Decide Course of Action on Lake Pollution," *Post-Standard*, March 1, 1952, 11.

87 "City, Geddes Plan Sewage Agreement," *Syracuse Herald-Journal*, March 5, 1952, 6.

88 "Syracuse Disposal Problem," *Herald-American*, March 23, 1952, 20.

89 "City to Finish Solvay Pipe Job Monday," *Syracuse Herald-Journal*, April 19, 1952, 16.

90 "To End the Pollution of Onondaga Lake," *Syracuse Herald-Journal*, March 12, 1952, 24.

91 "State Recommends 3 Classifications for Lake," *Post-Standard*, May 9, 1952, 8; "Pitts Criticizes Classification for Onondaga Lake," *Post-Standard*, May 10, 1952, 6.

92 "Geddes Board Drops City's Sewage Plan," *Syracuse Herald-Journal*, May 14, 1952, 4.

93 "City Loses Fight to Dump Sewage in Lakeland," *Post-Standard*, May 14, 1952, 6

94 "New Solvay Soda Plant Dedicated," *Syracuse Herald-Journal*, May 10, 1952, 3.

95 "Onondaga Lake's Mud Lock Slated for Restoration," *Post-Standard*, May 26, 1952, 5.

96 "Sportsmen Demand Clean Onondaga Lake," *Syracuse Herald-American*, June 7, 1952, 23; Red Hunter, "In the Sportsman's Corner," *Post-Standard*, June 22, 1952, 35.

97 "Stop Making Cesspool Out of Onondaga Lake, Pitt Tells Control Board," *Post-Standard*, June 28, 1952, 6.

98 "Mayor Seeks New City Sewage Disposal Plant," *Post-Standard*, October 12, 1952, 1, III.

99 "New 'Golden Age' Due for Lake," *Syracuse Herald-American*, October 19, 1952, 1, III.

100 "Pitts Lauded for Smoke Control Work," *Syracuse Herald-American*, October 19, 1952, 42.

101 "Mayor Proclaims Cleaner Air Week," *Post-Standard*, October 21, 1952, 12.

102 "2 Crews Speed Work on New Sewage Line," *Syracuse Herald-Journal*, March 22, 1953, 6.

103 "Mead Hits Democrats in Sludge Drive," *Syracuse Herald-American*, April 16, 1954, 10.

104 "Sludge Plan Working OK, Mead Says," *Syracuse Herald-Journal*, April 23, 1954, 10.

105 "Vast Sections of Lake to Permit Swimming," *Post-Standard*, April 12, 1953, 32.

106 "Mayor Promises Report by June 1 on Lake Pollution," *Post-Standard*, May 11, 1954, 6.

107 "Mead Submits Plan," *Post-Standard*, May 30, 1954, 19.

108 Grace Lewis, "At the Post," *Post-Standard*, May 22, 1953, 23.

109 "Before Common Council Approves Sewage Plan," *Syracuse Herald-Journal*, November 30, 1953, 33.

110 "Changes Urged in Plan 3 Sewage Proposal," *Post-Standard*, September 1, 1953, 6.

111 "Stop Water Pollution, State Warns," *Syracuse Herald-Journal*, May 7, 1955, 3.

112 "Steady Progress on Lake Pollution Problem Reported," *Post-Standard*, July 1, 1955, 24.

113 David Tymofy, "A Worthy Project" [Letter to the Editor], *Post-Standard*, March 13, 1954, 3.

114 "First, Railroad Curb Pollution," *Syracuse Herald-Journal*, August 7, 1957, 2.

115 "State Slates Hearings on Pollution," *Syracuse Herald-Journal*, August 22, 1957, 2.

116 "Liverpool Officials Miss Talks," *Syracuse Herald-Journal*, October 28, 1957, 2.

4. ECOLOGIES OF CONTAMINATION

1 Rachel Carson, *Silent Spring* (Boston: Houghton-Mifflin, 1962).

2 New York State Department of Environmental Conservation, *Baseline Ecological Risk Assessment (BERA)*, 2–3, December 2002, www.lakecleanup.com.

3 "City Issues Permit to Build Sewage Plant," *Post-Standard*, January 23, 1958; "5.5 Million Sewage Flow Units Scheduled to Be Advertised Soon," *Syracuse Herald-Journal*, February 24, 1958, 2; "Sewage Plant Bids Set," *Syracuse Herald-Journal*, February 26, 1958, 2.

4 "Commission Tours Pumping Station," *Syracuse Herald-Journal*, March 5, 1958, 31.

5 Mary Wilkinson, "Onondaga Lake: Upstate's Cinderella at Regatta Time," *Syracuse Herald-Journal*, June 15, 1958, 31.

6 "City Accepts $47,500 Grant," *Post-Standard*, January 19, 1960, 7; "Council to Get Pollution Survey Bill," *Post-Standard*, February 14, 1960, 8.

7 "Camillus to Take Action to End Creek Pollution," *Post Standard*, June 39, 1963, 22.

8 "2 Polluted Creeks Killing Game Fish," *Syracuse Herald-American*, October 4, 1959, 43.

9 "First Still Fighting War against Smog," *Post Standard*, February 13, 1964, 22.

10 "Kaylor to Speak on Air Pollution," *Post-Standard*, January 11, 1963, 7.

11 Jonathan M. Davidson and Joseph M. Norbeck, *An Interactive History of the Clean Air Act* (New York: Elsevier Science, 2011).

12 "McCarthy Urges Anti-Pollution Unit," *Post-Standard*, January 6, 1965, 6.

13 "In the Sportsman's Corner," *Post-Standard*, February 7, 1965, 36.

14 Liz Hannon, "Ask 'Open Sewer' Upgrading," *Post-Standard*, February 10, 1966, 17.

15 "John H. Mulroy," Wikipedia, https://en.wikipedia.org. Accessed August 1, 2019.

16 "Committee to Begin Study of Lake Pollution," *Post-Standard*, April 10, 1965, 7.

17 "Hails Lake Study," *Post-Standard*, April 17, 1968, 33.

18 Gene Goshorn, "Mulroy Offers 9-fold Lake Clean-up Plan," *Syracuse Herald-Journal*, April 18, 1965, 25.

19 Gene Goshorn, "Mulroy Criticized on Ash Dumping," *Syracuse Herald-Journal*, July 13, 1965, 2.

20 "Solvay Process Speakers Give Public 'Inside Look' at Company, *Syracuse Herald-American*, September 16, 1965, 23.

21 "34 Towns, Cities, Cited as Polluters," *Syracuse Herald-Journal*, August 19, 1965, 43.

22 Richard Long, "Pollution Programs Urged: Hanley Lauds Council Report," *Syracuse Herald-Journal*, March 24, 1966, 57.

23 "Study Finds Algae Increasing in Lake," *Syracuse Herald-Journal*, February 14, 1968, 21.

24 "Park Urged for Solvay Wastebeds," *Syracuse Herald-Journal*, March 25, 1966, 23.

25 "Slime Clogs Solvay's Creek Pipes," *Syracuse Herald-Journal*, April 1, 1966, 23.

26 "County to Run Sewers," *Syracuse Herald-Journal*, June 8, 1966, 49.

27 "Aid Rise Seen for Pollution End," *Post-Standard*, August 19, 1966, 6.

28 "Lake Clean-up Study Announced," *Syracuse Herald-Journal*, December 22, 1966, 4.

29 "Aid Rise Seen for Pollution End," *Post-Standard*, August 19, 1966, 6.

30 "Sewer Study Approved," *Post-Standard*, August 19, 1966, 6.

31 "Allied Chemical Offers S.U. $10,000 Pollution Grant," *Post-Standard*, October 25, 1966, 13.

32 "2 Industries to Stop Putting Waste in Lake," *Syracuse Herald-Journal*, December 23, 1966, 1.

33 "2 Firms to Halt Lake Dumping," *Syracuse Herald-American*, November 20, 1966, 18; "A Clean-up Timetable," *Syracuse Herald-Journal*, November 23, 1966, 31.

34 "$150000 More Approved for New Lake Park," *Syracuse Herald-Journal*, December 27, 1966, 19.

35 Dorothy Newer, "Pollution Cures Urged," *Post-Standard*, December 2, 1966, 13.

36 Don Casilio, "County to Run Sewers: It's a Step to End Pollution of the Lake," *Syracuse Herald-Journal*, June 8, 1966, 49.

37 "Ley Plant Work Ok'd," *Post-Standard*, August 2, 1966, 11.

38 Liz Mannon, "Onondaga Lake: 'A Stinking Mess,'" *Post-Standard*, June 20, 1967, 9.

39 Luther Bliven, "Connor Maps Pollution Fight," *Post-Standard*, June 20, 1967, 9.

40 Connie Myer, "Anti-Pollution Rules Conflict," *Post-Standard*, April 12, 1967, 11.

41 "Industry Called Air Pollution Scapegoat," *Syracuse Herald-Journal*, April 12, 1967, 43.

42 "Solvay Process Helps Clear the Air," *Syracuse Herald-American*, March 19, 1967, 24.

43 Gene Gosborn, "Solvay Process Eyes River Water," *Syracuse Herald-Journal*, March 2, 1967, 39.

44 "Mayor Issues Statement Explaining City's Position on Sewage Dispute," *Syracuse Herald-Journal*, March 11, 1968, 8.

45 "Lakeside Park to Rise on Reclaimed Land," *Post-Standard*, July 16, 1968, 6.

46 Connie Myer, "Sewage Plan Called Wasteful," *Post-Standard*, May 5, 1970, 12.

47 Dorothy Newer, "It's 'Back to Syracuse' for Allied Chemical," *Post-Standard*, January 22, 1970, 27.

48 M. James Campbell, "Allied Lists Program to Communicate," *Post-Standard*, February 2, 1970, 30.

49 Donald J. Lawless, "Residents Oppose Manlius Quarry," *Post-Standard*, June 4, 1970, 8.

50 Dorothy Newer, "Allied Faces U.S. Suit," *Post-Standard*, July 25, 1970, 1, 7.

51 Dorothy Newer, "Pollution Deadline Monday," *Post-Standard*, July 31, 1970, 13.

52 "Allied Submits Clean-up Plan," *Post-Standard*, August 4, 1970, 11.

53 "27 Million Okayed for Sewage Plant," *Syracuse Post Standard*, July 30, 1091, 26.

54 Frank Brieaddy, "Planners Favor Quarry Spread to Manlius," *Syracuse Post-Standard*, November 23, 1972, 6.

55 Paul Seminoff, Letter to the Editor, *Post-Standard*, April 5, 1972, 4.

56 "The Air You Breathe," *Syracuse Post-Standard*, March 1, 1972, 1, 4.

57 Paul Hornak, "Allied Quarry Plan Rejected by Manlius," *Syracuse Post-Standard*, August 22, 1973, 6.

58 "Anti-Pollution Bonds Planned," *Syracuse Post-Standard*, February 8, 1974, 10.

59 Dorothy Newer, "Allied Gives Rosy Forecast," *Syracuse Post-Standard*, April 30, 1974, 21.

60 Dorothy Newer, "Say Tests Show Allied Chemical Pollutants," *Syracuse Post-Standard*, August 30, 2016, 6.

61 "Bond Okay Aids Firms," *Post-Standard*, November 22, 1974, 11.

62 "Sierra Club Criticizes Sewage Plant Design," *Post-Standard*, February 4, 1974, 7.

63 Jonas A. Gylys, "Opposes Zone Change for Allied," *Post-Standard*, June 12, 1974, 8.

64 "Chemical Firm Dips in Earnings," *Syracuse Herald-American*, October 12, 1975, 43.

65 "Fish Get Allied Off Hook," *Syracuse Herald-Journal*, November 6, 1976, 13.

66 "Allied Chemical Indicted," *Post-Standard*, May 8, 1976, 2.

67 Michael Kelly, "Onondaga Lakes Lives Again: Elaborate Park Planned," *Herald American*, June 26, 1977, 59.

68 Dan Padovano, "Landfill Fight: Quarry Hearing Tomorrow," *Syracuse Herald-Journal*, June 9, 1977, 41.

69 "From 'Paradise' to 'Garbage Pit,'" *Syracuse Herald-Journal*, January 26, 1977, 19.

70 Dan Padovano and Ezra Greenhouse, "County to Defy No-Dumping Rule at Clay Landfill," *Syracuse Herald-Journal*, January 3, 1977, 17; Michael Kelly, "Legislature Seeking Solid Waste Solution," *Herald-American*, January 9, 1077, 51.

71 "Allied, Manlius to Meet Tonight," *Syracuse Post-Standard*, April 6, 1978, 18.

72 "Allied Brine Spill Leaves Creek, Fisherman's Days Empty," July 27, 1979, 6.

73 "Pipe Leak Alarm Sought," *Syracuse Herald-Journal*, August 2, 1979, 33.

74 "Allied Balks at J'ville Quarry for Landfill," *Syracuse Herald-Journal*, April 19, 1979, 49.

75 "Allied: No Plans to Close Plant," *Syracuse Herald-Journal*, April 24, 1979, 21.

76 Carol L. Cleveland, "Allied Cleans Its Act—And the Air," *Syracuse Post-Standard*, June 5, 1980, 11.

77 Janet Meyers, "Manlius Board Can't Win in Allied Case," *Syracuse Post Standard*, March 13, 1980, 5.

78 Barbara Shelly, "100 Years Old: Solvay Process to Stage Centennial Celebration Today," *Syracuse Post-Standard*, September 12, 1981, A7.

79 Michael Kelly, "Revival: Lakes Recreational Potential Worth the Effort, Report Says," *Syracuse Herald-Journal*, March 18, 1983, D3.

80 Michael Kelly, "County Disputes Charge that It Is Fouling Onondaga Lake," *Syracuse Herald-Journal*, July 19, 1984, B3.

81 Bill Crozler, "Allied Chemical Sells Part of Solvay Plant to Local Firm," *Syracuse Herald-Journal*, December 29, 1984, A3.

82 Andrew Albert, "Cancer Rate Linked to Wastes," *Syracuse Herald-Dispatch*, January 22, 1981, D3.

83 Elizabeth Kolbert, "The Talk of Syracuse; New York's Old 'Salt City' Struggling to Overcome Setbacks," *New York Times*, June 16, 1986, www.nytimes.com. Accessed November 8, 2016.

84 "Allied's Plan to Leave Shocks Syracuse," *New York Times*, April 29, 1985, 3.

85 Robert W. Andrews, "From Clean Lake to Witches Brew: Onondaga Lake: A Paradise Lost?" *Post-Standard*, October 14, 1985, A1, A10.

86 Ibid., A10–A11.

87 Robert W. Andrews, "While Politicians Talked, the Lake Began to Die: Onondaga Lake: A Paradise Lost?" *Post-Standard*, October 16, 1985, A1, A6, A7.

88 Andrews, "From Clean Lake to Witches Brew," A10–A11.

89 Ibid., A11.

90 Robert W. Andrews, "Allied's Departure Sparks Optimism—and a Call for Patience: Onondaga Lake: A Paradise Lost?" *Post-Standard*, October 17, A8–A9.

91 Robert W. Andrews, "DEC Sets Agenda for Allied Clean-up," *Post-Standard*, October 29, 1985, A1.

92 "Allied's Closings Similar," *Post Standard*, November 4, 1985, A9.

93 Robert W. Andrews, "Allied Reveals Plan to Sell Its Wastebeds," *Post-Standard*, November 20, 1985, B1.

94 Tom Rose, "Construction Rises in Western Towns," *Post-Standard*, February 3, 1986, D3.

95 Robert W. Andrews, "Legislature Passes Watered Down Version of Lake Study," *Post-Standard*, February 4, 1986, B1.

96 Robert W. Andrews, "Mulroy Seeks Answer to Mess Problems," *Post-Standard*, March 18, 1986, B3.

97 Mike Grogan, "Trash-Burning Plant Proposal to Be Studied," *Post-Standard*, August 12, 1986, B1.

98 Robert W. Andrews, "Power Project Proposed for Allied Site," *Post-Standard*, August 12, 1986, B1.

99 Jonathan Bor, "Power Plant May Face Stricter Emissions Test," *Post-Standard*, September 18, 1987, A2.

100 James T. Mulder, "PSC Asked to Act on Two Power Projects," *Syracuse Herald-Journal*, December 3, 1986, B9; Kenneth Tompkins, "DEC May Require Lengthier Study from Hydra-Co," *Post-Standard*, December 3, 1986, B17.

101 James T. Mulder, "Salt City Cogeneration Stalled by PSC," *Syracuse Herald-Journal*, December 18, 1986, E7.

102 Robert W. Andrews, "County: Dump Sludge on Shore," *Post-Standard*, July 4, 1986, A2.

103 Robert W. Andrews and Jonathan Bor, "Allied to Examine Extent of Pollution from Wastebeds," *Post-Standard*, November 1, 1986, B1.

104 Robert W. Andrews, "Lake Still Lacks Adequate Level of Oxygen, Test Shows," *Post-Standard*, November 18, 1986, B1.

105 William Kates, "Can It Be Saved? Onondaga Lake Still Suffers Sting of Pollutants," *Syracuse Herald-American*, July 19, 1987, F8.

106 The Mission was established by the French in the early seventeenth century with the purpose of converting what the colonists then referred to as "the Iroquois." The Iroquois, apparently less than enthusiastic about its presence, threatened to attack it. In response, the members of the Mission vacated it. The museum was closed in 2013, after years of declining attendance, and the desire of Onondaga County to use the facility for office space. ("Sainte Marie among the Iroquois," Wikipedia, https://en.wikipedia.org. Accessed December 13, 2021).

107 "Lake Lovers Hope for a Comeback," *Syracuse Herald-American*, August 9, 1987, A12.

108 "Study Finds Worsening Mercury Contamination in Lake," *Syracuse Herald-American*, September 9, 1987, A10.

109 Jonathan Bor, "Power Plant Would Bring Jobs, Pollution Back to Allied Site," *Post-Standard*, September 17, 1987, A2.

110 Gary Gerew, "Neighbors, Industry Clash Over Hyrda-Co Plan," *Syracuse Herald-Journal*, September 10, 1987, B1.

111 "Trucking of Sludge a 'Waste of Money,'" *Post-Standard*, October 21, A9.

112 Jonathan Salant, "Mayor Young Goes to Washington," *Syracuse Herald-Journal*, October 22, 1987, B7.

113 Robert W. Andrews, "Caustics Dumped in Brook," *Post-Standard*, January 8, 1988, B1.

114 Kenneth Tompkins, "Company Arrived in 1979 To Be a 'Good Neighbor,'" *Post-Standard*, June 29, 1988, A7.

115 "Pollution Prompts Closure, Repairs," *Post-Standard*, June 29, 1988, A7.

116 "County Misses Pollution Deadline," *Syracuse Herald-Journal*, July 1, 1988, B1.

117 "LCP Could Face Criminal Charges in State Probe," *Syracuse Herald-Journal*, July 14, 1988, A12.

118 "LCP Lays Off Most Workers as Talks Proceed with DEC," *Post-Standard*, July 14, 1988, A6.

119 "Legislators' Vote Preserves $18 Million in Sewer Aid," *Post-Standard*, August 13, 1988, A7.

120 "DEC: Where Have All the Oil Spills Gone," *Post-Standard*, November 11, 1988, B1.

121 Marie Morelli, "Pollution Plagues Oil City," *Syracuse Herald-Journal*, June, 20, 1989, B1.

122 "LCP Chemical Likely Superfund Site," *Post-Standard*, April 28, 1989, B1.

123 "LCP Chemical Site a Likely Addition to Superfund List," *Post-Standard*, April 28, 1989, B1.

124 "Mall Will Float on Soft, Mucky Soil," *Syracuse Herald-Journal*, July 4, 1989, B1.

125 "Lawmakers Must Correct Polluted Waters," *Syracuse Herald-American*, January 29, 1989, E9.

126 Marie Morelli, "Senator Demands State Legal Action against Allied," *Syracuse Herald-Journal*, April 27, 1989, C1.

127 Mike Fish, "Committee Gives OK for Funds for Lake Work," *Post-Standard*, July 21, 1989, B1.

128 Esther Gross, "Bio-Gro Offering to Solve County Sludge Problem, Rest Fears of Camillus Residents Near Waste Beds," *Syracuse Herald Journal*, January 25, 1989, B5E.

129 Glenn Coin, "The Thirty Year Old Lawsuit that Cleaned Up Onondaga Lake Is about to End," October 7, 2019, www.syracuse.com.

130 *State of New York v. Allied-Signal, Inc.*, No. 3:89-cv-00815 (N.D.N.Y. June 27, 1989).

5. A THIRD WAVE OF ENVIRONMENTALISM

1 William R. Jordan and George M. Lubick, *Making Nature Whole: A History of Ecological Restoration* (Washington, DC: Island Press, 2011), 1.

2 Aldo Leopold, *A Sand County Almanac: And Sketches Here and There* (New York: Oxford University Press, 2020), William R. Jordan III, *The Sunflower Forest* (Oakland: University of California Press, 2003).

3 James Hamilton, "Careers in Environmental Remediation," U.S. Bureau of Labor Statistics, www.bls.gov. Accessed November 17, 2021.

4 See, for example, Leo van Velzen, ed., *Environmental Remediation and Restoration of Contaminated Nuclear and NORM Sites* (Cambridge, England: Waltham

Publishing, 2015), Organization for Economic Co-operation and Development, International Atomic Energy Agency, *Environmental Remediation of Uranium Production Facilities* (Paris: OECD Publishing, 2002), National Research Council, *Science and Technology for Environmental Clean-up at Hanford* (Washington, DC: National Academy Press, 2001).

5 Naveed Ahmad, et al., "Stakeholders' Perspective on Strategies to Promote Contaminated Site Remediation and Brownfield Redevelopment in Developing Countries: Empirical Evidence from Pakistan," *Environmental Science and Pollution Research International* 27, no. 1 (2020): 14614–14633; Timothy J. Dixon, *Sustainable Brownfield Regeneration: Liveable Places from Problem Spaces* (Oxford: Blackwell, 2017); Richard T. Melstrom and Rose Mohammadi, "Residential Mobility, Brownfield Remediation and Environmental Gentrification in Chicago," *Land Economics* (October 2021). 60520-0077R1.

6 Stuart K. Allison, *Ecological Restoration and Environmental Change: Renewing Damaged Ecosystems* (London: Routledge, 2012), 37.

7 Cindy Amrhein, *A History of Native American Land Rights in Upstate New York* (Mt. Pleasant, SC: Arcadia Publishing, 2008).

8 Joseph Alexiou, *Gowanus: Brooklyn's Curious Canal* (New York: New York University Press, 2015), 353.

9 Eric Higgs, *Nature by Design: People, Natural Process, and Ecological Restoration* (Cambridge, MA: The MIT Press, 2003), 256.

10 Jonathan Salant, "Progress on Cleaning Up Lake Praised: Better Water Quality, Lake Developments Cited by Officials," *Post-Standard*, January 6, 1990, B3.

11 Jonathan Salant, "Syracuse Winner in Clean Air Fight," *Syracuse Herald-American*, A18.

12 "Toxic Soil Piles Up at Carousel Site: Reports Continue of Safety Violations at Mall Site," *Syracuse Herald-Journal*, July 17, 1990, A6.

13 "Anglers Ignore Onondaga Lake's Diverse Fishes," *Post-Standard*, June 20, 1991, D1, D4.

14 Cindy W. Rodriguez, "Power Plant in Geddes is Back on Track," *Syracuse Herald-Journal*, December 10, 1991, B6.

15 "Scientists Track Salty Water in Creek to NiMo's Pipes," *Post-Standard*, July 23, 1991, A1, A4.

16 Editorial, "Salty Lake: Its Reputation Predates the First Jesuit Sip," *Post-Standard*, December 7, 1991, A10.

17 Timothy P. Mulvey, "Humans Gave Waters a Bad Rep," *Post-Standard*, December 26, 1991, A9.

18 Robert W. Andrews, "Lake's 'Tar' Is New Hazard: State Searching for Source of Pollution Patch," *Post-Standard*, July 12, 1991, B3.

19 U.S Congress, Senate Committee on Environment and Public Works, *Progress on the Onondaga Lake Management Conference*, Washington DC: GPO, 102[d] Cong., 2d sess., December 9, 1992, 49–53.

20 Ibid., 19.

21 Margaret LeBrun, "Lake Proves a Murky Topic for Teens: Youth Roundtable Hears about Plans to Clean Up the Polluted Body of Water," *Syracuse Herald-Journal*, October 31, 1994, B3.

22 Timothy P. Mulvey, "Repeating Past Mistakes with Lake Will Only Cost More in the Long Run," *Post-Standard*, July 10, 1994, A7.

23 Lynn Davis, "Onondaga Lake a Tourist Lure? County Could Attract Morbid Curiosity Seekers," *Syracuse Herald-Journal*, December 29, 1994, A8.

24 Lillian Abbott Pfohl, "Smaller Companies Take Root in Solvay," *Post-Standard*, April 21, 1995, A10.

25 Mark Weiner, "Lake Group Fishes for Pollution Solution," *Post-Standard*, October 30, 1996, C1.

26 "Political Promises for the Environment," *Syracuse Herald-Journal*, October 28, 1997, A12.

27 Mark Weiner, "The Renewal of Onondaga Lake," *Post-Standard*, October 27, 1997, B-8.

28 J. Michael Kelly, "Lake's Fish Population Is Diversifying," *Syracuse Herald-Journal*, June 25, 1999, A12.

29 "Company Promises to Finish Its Clean-up within Seven Years," *Syracuse Herald-American*, November 28, 1999, A12.

30 "The History of Honeywell," www.honeywell.com. Accessed December 9, 2021.

31 "Company Promises to Finish," A12.

32 Mark Weiner, "Scientists Back Change in Lake Clean-up Plan," *Syracuse Herald-Journal*, April 9, 1999, B3.

33 Gloria Wright, "City Moves to Collect Delinquent Penalties," *Syracuse Herald-Journal*, January 27, 1999, C1.

34 Mark Weiner, "LCP Plant Clean-up Planned," *Herald-Journal*, February 17, 2000, C1.

35 Mark Wiener, "Drilling for Toxics: 700 Samples Well Be Taken from Lake Bottom," *Herald-American: The Post Standard*, July 23, 2000, A1.

36 Mark Weiner, "Suits Raises Lake-Clean-up Question," *Post-Standard*, December 6, 2000.

37 "Official Defends Company's Effort," *Post-Standard*, November 11, 2001, B2.

38 Ibid.; Mark Weiner, "Company Plans to Put Tar Beds to Bed," *Post-Standard*, February 6, 2002, B2.

39 "Part of Mall Parking Lot on Hazardous Waste Site," *Post-Standard*, October 22, 2002, A9.

40 "Public Clean-up Work Moves Ahead; Private Work Falters," *Post-Standard*, June 9, 2002, A2.

41 Mark Weiner, "Polluter Spreads Blame," *Post-Standard*, A1, A4.

42 New York State Department of Environmental Conservation, "Remedial Investigation Report" [RI], December 2002, 11, www.lakecleanup.com.

43 Ibid., 12.

44 Ibid., Chap. 4, 11.

45 Ibid.

46 Ibid., 14.
47 Ibid., 14.
48 Ibid., 14.
49 Ibid., 18.
50 Ibid., 19.
51 Ibid., Executive Summary, 2.
52 Ibid., 13.
53 New York State Department of Environmental Conservation, "Human Health Risk Assessment" [HHRA], Chap. 2, 57, www.lakecleanup.com.
54 RI, Executive Summary, 3.
55 Ibid., 4
56 Ibid., 6–7.
57 Ibid., 15.
58 Ibid., 16.
59 Ibid., 20, 21.
60 Ibid., 17.
61 Ibid., 18–20.
62 HHRA, Chap. 4, 3.
63 Mark Weiner, "Honeywell Files Clean-up Plan with DEC," *Post-Standard*, B-5.
64 New York State Department of Environmental Conservation [NYSDEC], "Onondaga Lake Bottom Subsite of the Onondaga Lake Superfund Site: Proposed Plan," November 29, 2004, 27–28. www.dec.ny.gov.
65 Ibid., 29.
66 Ibid., 30.
67 Ibid., 31.
68 Ibid., 31.
69 Ibid., 31–32.
70 Ibid., 32.
71 Ibid., 1.
72 Mark Weiner, "State Rejects Industry Plan to Clean Up Lake," *Post-Standard*, January 4, 2004, A1, A12.
73 Mark Weiner, "Nonhazardous Material Would Go to Camillus, Geddes," *Post-Standard*, July 30, 2004, A11.
74 Pam Greene, "DEC Offers Lake Clean-up Plan for Review," *Post-Standard*, November 30, 2004, B-5.
75 Editorial, "Lake Bottom Line," *Post-Standard*, November 24, 2004, A14.
76 Mark Weiner, "Input Sought on Lake Clean-up," *Post-Standard*, December 29, 2004, B2.
77 NYSDEC, "Onondaga Lake Bottom Subsite," 48–54.
78 Mark Weiner, "More Debate Sought on Lake Clean-up Proposal," *Post-Standard*, December 4, 2004, B3.
79 Mark Weiner, "Onondaga Lake Clean-up Topic of Honeywell Poll," *Post-Standard*, December 13, 2004, B3.

80 Mark Weiner, "Brine Fields May Have Moved," *Post-Standard*, July 23, 2004, B1, B2.

81 Mark Weiner, "Camillus Frets Over Onondaga Lake Clean-up Plan," *Post-Standard*, February 17, 2005, B-6.

82 Mark Weiner, "$20 Million Plant to Aid Clean-up," *Post-Standard*, April 8, 2005, B2.

83 Editorial, "Onondaga Lake on the Mend," *Post-Standard*, August 24, 2005, A10.

84 Mark Weiner, "Major Source of Mercury Contamination Cut Off," *Post Standard*, September 23, 2005, B5.

85 Mark Weiner, "Lake Problems Remain, Critics Say," *Post-Standard*, November 3, 2005, B2.

86 Pam Greene, "ESF Scientists Show Progress in Onondaga Lake Clean-up," *Post-Standard*, October 5, 2005, B3.

87 Mark Weiner, "Barrier Will Keep Out Tainted Water," *Post-Standard*, October 12, 2005, B2.

88 Mark Weiner, "Progress in Lake Clean-up," *Post-Standard*, November 17, 2005, B1.

89 Mark Weiner, "Buried Mercury Near Lake Criticized," *Post-Standard*, December 4, 2005, B3.

90 Mark Weiner, "New Lake Study to Come," *Post-Standard*, May 15, 2005, B3.

91 Mark Weiner, "Onondaga Lake Clean-up Decision by State DEC Due Today," *Post-Standard*, July 1, 2005, B-6.

92 "Shareholders Say Signs around the Lake Unnecessary," *Post-Standard*, April 25, 2006, B3.

93 Mark Weiner, "State Okays Lake Clean-up Proposal," *Post-Standard*, July 2, 2005, B2.

94 "Coming in 2006: Onondaga Lake Clean-up," *Post-Standard*, December 26, 2005, B2.

95 Letter to Alma L. Lowry from the United States Environmental Protection Agency, "Re: Dismissal of Administrative Complaint 3R-04-R2. Midland Regional Treatment Facility, Syracuse, New York," March 18, 2005, 2, www.onondagacreek.org.

96 Ibid., 6–7.

97 The Partnership for Onondaga Creek, "Position Paper on the Midland RTF—An Opportunity for a Better Solution," November 2, 2012, www.onondagacreek.org.

98 *City of Syracuse v. Onondaga County*, 464 F.3d 297 (2006).

99 Mark Weiner, "New Source of Toxins Found at Lake Site," *Post-Standard*, June 4, 2006, 1, 8.

100 Mark Weiner, "Willows Might Reduce Lake Pollution," *Post-Standard*, September 15, 2006, A1.

101 *Cayuga Indian Nation of New York v. Pataki*, 165 F. Supp. 2d 266 (N.D.N.Y 2001).

102 "Onondaga Claim 4,000 Square Miles of New York," *Capital*, March 12, 2005, A3.

103 *Cayuga Indian Nation of New York v. Pataki*, 413 F.3d 266 (2d Cir. 2005).

104 Sidney Hill, "Changing the Rules: The Courts Blocked Land Claims for Years: Now They Say Indian Nations Waited Too Long," *Post-Standard*, July 3, 2005, C3.

105 *Onondaga Nation v. State of NY*, 10–4273 (2d Cir. 2012)

106 Delen Goldberg, "Goodall Makes Appeal to Save Onondaga Lake," *Post-Standard*, September 20, 2006, A4.

107 Mark Weiner, "Little Known of Cleanup," *Post-Standard*, August 14, 2006, B-5.

108 Wendy Gonyea, interviewed by author, March 11, 2022.

6. A SACRED LAKE, A RECREATIONAL RESOURCE

1 Delen Goldberg, "How to Detoxify a Lake. Step One: Build a Barrier Wall," *Post-Standard*, October 15, 2006, A1–A2.

2 Philip P. Arnold, "Onondaga-Our Lake: Learn to Think of It in a New Way," *Post-Standard*, October 22, 2006, E3.

3 Margaret Harrigan, Letter to the Editor, "Chargers Endorse Consent Decree for Onondaga Lake," *Post-Standard*, November 14, 2006, A11.

4 Samuel L. Sage, Letter to the Editor, "A Sparkling Future: Ultimately, the Public Controls the Fate of Onondaga Lake," *Post-Standard*, A11.

5 Delen Goldberg, "Newest Lake Clean-up Hits Snag: State, Honeywell Get 22 More Months to Work on Tributaries," *Post-Standard*, January 17, 2007, A1, A9.

6 John McAuliffe, Letter to the Editor, "Onondaga Lake Clean-up Actually Ahead of Schedule," January 21, 2007, E3.

7 "Pirro's Progress," *Post-Standard*, February 24, 2007, A4.

8 "Farewell: Nicholas Pirro, Onondaga County Executive, 1988–2007," *Post-Standard*, December 26, 2007, A4.

9 Robert W. Andrews, "Accelerating the Future," *Post-Standard*, August 1, 2007, 36–40.

10 Charley Hannagan, "Stem to Stern: How Anheuser-Busch Brewery in Lysander Wrings Savings from Every State of Production," *Post-Standard*, February 11, 2008, I-6.

11 John Mariani, "Trail Plan Comes Naturally: Project Would Extend Onondaga Lake Canalway Path," *Post-Standard*, July 12, 2007, B1.

12 "Tree Planting Marks Step in Restoring Lake's Habitat," *Post-Standard*, October 19, 2007, B1.

13 Meghan Rubado, "New Trees Mark New Beginning for Former Hazardous Site," *Post-Standard*, November 1, 2007, 12.

14 Meghan Rubado, "Lake Lives Again: Onondaga's Lake's Water Quality Best in 40 Years," *Post-Standard*, November 16, 2007, A1, A4.

15 Dick Case, "It's Not Easy Restoring Onondaga Lake Green Space," *Post-Standard*, November 25, 2007, B-6.

16 "Steel Sheets Will Go Up Along the Lake's West Shore," *Post-Standard*, December 18, 2007, B2.

17 Mark Weiner, "Scientists Marvel at 'Spectacular Progress' in Onondaga Lake," *Post-Standard*, September 28, 2008, A1, A6.

18 Editorial, "A Cleaner Lake," *Post-Standard*, October 6, 2008, A8.

19 "Lake Passes Chemistry Test," *Post-Standard*, September 28, 2008, A6.

20 Delen Goldberg, "Newest Mercury Is Most Harmful," *Post-Standard*, December 1, 2008, B2.

21 John Mariani, "To Spur Willow's Growth," *Post-Standard*, July 1, 2008, B1.

22 Dick Case, "Bringing Back Life to Wastebeds Wasteland," *Post-Standard*, July 27, 2008, B3.

23 Delen Goldberg, "The Earth Is Falling in Tully Valley," *Post-Standard*, November 2, 2008, B1, B3.

24 Delen Goldberg, "Comments Split on DEC Clean-up of Waterways," *Post-Standard*, December 11, 2008, B3.

25 David M. Stott, Letter to the Editor, "Lake Clean-up Should Be Done Right This Time," *Post-Dispatch*, December 30, 2008, A7.

26 John Mariani, "Fisheries Board Sets First Session," *Post-Standard*, February 20, 2009, B3.

27 *Atlantic States Legal Foundation v. Onondaga County*, U.S. District Court, Northern District of New York, 88-CV-066, November 16, 2009.

28 Christopher Kloss and Crystal Calarusse, *Rooftops to Rivers: Green Strategies for Controlling Stormwater and Combined Sewer Overflows* (New York: Natural Resources Defense Council, 2006), 7–12.

29 Glenn Coin, "Save the Rain Program Wins Environmental Excellence Award for Onondaga County," Syracuse.com, March 22, 2019, www.syracuse.com.

30 "Featured Projects," Save the Rain, website, https://savetherain.us. Accessed July 7, 2021.

31 Tim Knauss, "Before They Dredge the Lake . . ." *Post-Standard*, January 13, 2010, A3.

32 Tim Knauss, "Onondaga Lake Clean-up Plan Angers Some Camillus Neighbors," *Post-Standard*, January 15, 2010, A4.

33 Joe St. Louis, Letter to the Editor, "Don't Put Dumpsite Near School and Park," *Post-Standard*, January 27, 2010, A13.

34 John Mariani, "Camillus Says 'No' to Lake Toxins," *Post-Standard*, January 28, 2010, A14.

35 John Stith, "Another Idea for Camillus Lake Beds," *Post-Standard*, February 14, 2010, B3.

36 Richard Lauricella, Sarah Marinelli, Letters to the Editor, "Camillus Wastebeds," "Project Will Cause Health Problems and Damage Image of Town"; "Questions of Whether Honeywell Will Keep Its Word on Wastebeds," *Post-Standard*, March 6, 2010, A11.

37 Colleen Fesinger, Letter to the Editor, "Secrecy about Dredge Plan Hard to Research," *Post-Standard*, March 22, 2010, A11.

38 John Stith, "Making a Difference," *Post-Standard*, April 1, 2010, 3.

39 John Stith, "Wastebed Report Is 'Propaganda,' Critics Say," *Post-Standard*, April 12, 2010, A4.

40 John Stith, "Mounds Lifeless Despite Seeding," *Post-Standard*, May 29, 2010, A3. Likely, Gdula was referring to calcium chloride, which is the main byproduct of the Solvay process. While some calcium carbonate existed in the wastebeds, it was not their main constituent.

41 Honeywell Advertisement, "A Healthier Onondaga Lake," *Post-Standard*, May 30, 2010, A11.

42 Honeywell Advertisement, "Onondaga Lake Is on the Move Again," *Post-Standard*, July 8, 2010, A2.

43 Honey Advertisement, "Attracting Wildlife," *Post-Standard*, June 13, 2010, A6.

44 John Stith, "Talking Point: Wastebed 12 Study," *Post-Standard*, July 8, 2010, A4.

45 Mark Weiner, "1970: Don't Eat the Fish; 1999: Go Ahead and East Some Fish; 2007: Don't Eat Certain Fish; 2010: Don't Eat Almost Any Fish," *Post-Standard*, June 13, 2010, B1.

46 John Stith, "EPA Finds Sediment Disposal Poses No 'Significant Health Risks,'" *Post-Standard*, June 24, 2010, A2.

47 Alex Abda, Letter to the Editor, "Facts Substantiate Fears about Honeywell Clean-up Plan," *Post-Standard*, June 25, 2010, A11.

48 John Stith, "Honeywell May Not Need Local Approval," *Post-Standard*, July 23, 2010, A1.

49 John Stith, "Clean-up: Camillus Permits Not Needed," *Post-Standard*, August 13, 2010, A3.

50 John Stith, "Honeywell Hopes Rye Will Spur Greenery Growth," *Post-Standard*, July 5, 2010, A7.

51 Mark Weiner, "Waves of Support for Local Control of Lake Project," *Post-Standard*, July 6, 2010, A5.

52 Sean Kirst, "Effler: Onondaga Lake's Truthteller," *Post-Standard*, July 7, 2010, 2.

53 John Stith, "Camillus Man Asks Officials to Test Clean-up Technology," *Post-Standard*, July 15, 2010, 10.

54 John Stith, "Wastebed Preparations Continue," *Post-Standard*, September 16, 2010, A8.

55 "Park and Trails Will Grow in Chemical Wasteland," *Post-Standard*, November 17, 2010, A1, A3.

56 Clyde Ohl, "Willows Look Like Excuse to Avoid Costs," *Post-Standard*, November 18, 2010, A15.

57 Andy Mager, "Onondaga Lake Vision," *Post-Standard*, December 12, 2010, A11.

58 Teri Weaver, "Developer Tries His Luck with Two Racino Sites," *Post-Standard*, March 4, 2011, A1–A11.

59 Advertisement, "Onondaga Lake Bottom Dredging to Commence 2012: Preventing Toxic Pollution into Onondaga Lake Is Top Priority for 2011," *Post-Standard*, March 24, 2011, 12.

60 "Local Control," *Post-Standard*, April 8, 2011, A14.

61 David Figura, "Anglers Struggling to Catch Bass in Onondaga," *Post-Standard*, July 24, 2011, C6.

62 Rick Moriarty, "The Hidden Relics of Onondaga Lake," *Post-Standard*, August 14, 2011, B1.

63 David Figura, "Anglers Struggling to Catch Bass in Onondaga," *Post-Standard*, July 24, 2011, C6.

64 Rick Moriarty, "Meet Marlin, Here to Clean Up Your Lake," *Post-Standard*, April 21, 2012, A3, A6.

65 Rick Moriarty, "Event Makes the Start of Onondaga Lake Clean-up," *Post-Standard*, May 31, 2012, A4.

66 Paul Riede, "Dredging Enters Second Week," *Post-Standard*, August 6, 2012, A3.

67 Editorial, "A Means to an End, Focus on Sewers Leads to Cleaner Onondaga Lake," *Post-Standard*, September 25, 2012, A14.

68 Paul Riede, "Bad Smells Halt Lake Dredging," *Post-Standard*, October 2, 2012, A1, A13.

69 TenCate Geosynthetics Americas, https://www.tencategeo.us. Accessed August 9, 2021.

70 Paul Riede, "Odor in Check, Lake Dredging Resumes Today," *Post-Standard*, October 11, 2012, A3, A9.

71 Editorial, *Post-Standard*, October 9, 2012, A12.

72 Dick Case, "Progress Made Over Waste of Yesteryear," *Post-Standard*, November 4, 2012, B1, B3.

73 Paul Riede, "SU Survey Looks at Hopes for Lake," *Post-Standard*, November 18, 2012, B3.

74 Paul Riede, "Trail Work Slow but Steady," *Post-Standard*, December 3, 2012, A3.

75 Juliana Pignataro, "Residents: Something Stinks," *Post-Standard*, February 1, 2013, A3.

76 Camillus Clean Air Coalition, "Neighbors Are Convinced: Camillus Waste Site Stinks," *Post-Standard*, A21.

77 Minnich and Scotto, Inc., "Air Contaminant Exposure in Residents of the Town of Camillus from Honeywell's Sediment Containment Facility," April 8, 2013. Email attachment to author, August 3, 2021.

78 Minnich and Scotto, 4–3.

79 Minnich and Scotto, 5-1-5-17.

80 Minnich and Scotto, Inc., May 8, 2013. "Response to DEC Review of Minnich and Scotto's April 8, 2013 Report: 'Air Contaminant Exposure to Residents of the Town of Camillus from Honeywell's Sediment Treatment and Containment Facility,'" Email attachment to author, August 3, 2021.

81 Glenn Coin, "Lawyer for Lake Clean-up Group May Be Unfit, Judge Says," *Post-Standard*, August 15, 2013, 4.

82 Timothy Minnich, interviewed by author, August 3, 2021.

83 Glenn Coin, "Mercury's Falling in an Onondaga Lake Success Story," *Post-Standard*, October 13, 2013, A1, A8.

84 Tadodaho Sidney Hill, "An Onondaga Lake Success Story?" *Post-Standard*, October 20, 2013, E1, E4.

85 Charles Driscoll, David Matthews, and Steven Effler, "It's True: Onondaga Lakes Fish Mercury Levels Are Falling," *Post-Standard*, November 10, 2013, E1.

86 David Matthews, interviewed by author, September 22, 2021.

87 David Figura, "Audubon New York Hosting 'Birds of Onondaga Lake,'" *Post-Standard*, March 16, 2014, C11.

88 Michelle Breidenbach, "7 Questions: How to Build an Amphitheater on a Superfund Site," *Post-Standard*, March 27, 2014, A7.

89 Glenn Coin, "County Officials Say: Onondaga Lake Now Cleaner than Some Finger Lakes," *Post-Standard*, April 11, 2014, 1.

90 Glenn Coin, "Honeywell Reports 'Terrific Progress,' in Lake Clean-up Effort," *Post-Standard*, April 22, 2014, A4.

91 Glenn Coin, "Onondaga Nation Doesn't Participate in Lake Trail Extension Grand Opening," *Post-Standard*, May 22, 2014, 2.

92 Glenn Coin, "New York DEC: Honeywell's Clean-up Reports Aren't Public," *Post-Standard*, May 22, 2014, 6.

93 Glenn Coin, "Lawyer: Onondaga Lake Clean-up Is Poisoning Camillus Residents," *Post-Standard*, May 29, 2014, A3.

94 "Lake Dredging Resumes," *Burlington North Carolina Times News*, June 8, 2014, C8.

95 Glenn Coin, "Amphitheater Site Would Be 'as Safe as a Green Field,' County Says," *Post-Standard*, June 12, 2014, D1.

96 Gonyea interview.

97 David Figura, "Are Fish Still in Onondaga Lake?" *Post-Standard*, June 29, 2014, C11.

98 Glenn Coin, "Onondaga Lake is Cleaner, but Native Species Thrive," *Post-Standard*, September 29, 2014, A1, A4.

99 Glenn Coin, "Boiling Mud," *Post-Standard*, October 14, 2014, A1, A4.

100 Marnie Eisenstadt, "Late Vote Switch Gives Green Light to Amphitheater," *Post-Standard*, November 4, 2014, A1, A6.

101 Glenn Coin, "Dredging is Over, Criticism, Isn't," *Post-Standard*, November 4, 2014, A3, A4.

102 Glenn Coin, "Camillus Residents: Dredging of Lake Made Us Sick," *Post-Standard*, April 9, 2015, A3, A4.

103 Tim Knauss, "Mahoney Fights Scrapyard Near the Lake, but Not Because of Her Ex-Chief of Staff," *the Post-Standard*, June 7, 2015, A10.

104 David Figura, "State Plans to Pump $1.4 Million into 4 CNY Lakes," *Post-Standard*, October 5, 2015, A1, A4.

105 Glenn Coin, "Honeywell Receives Audubon's Highest Award for Lake Clean-up," *Post-Journal*, November 7, 2015, A1.

106 Chris Baker, "And They Call Live . . ." *Post-Standard*, July 23, 2015, A1, A8.

107 Glenn Coin, "In Mahoney's Budget: $300,000 to Study Onondaga Lake Beach," *Post-Standard*, A1.

108 Glenn Coin, "Lake Toxins Uncapped," *Post-Standard*, January 31, 2016, A5, A6.

CONCLUSION

1 *Bartlett v. Honeywell Int'l Inc.*, 737 Fed Appx, 550, May 9, 2017.

2 *Bartlett v. Honeywell*, 551.

3 "4.5 Million Gallons of Sewage Spews into Onondaga Lake as Pipe Bursts from Rain," Syracuse.com, October 31, 2017, www.syracuse.com; "For 90 Million Gallons of Sewage Spills into Onondaga Lake, County Fined $5,300," Syracuse.com, July 10, 2019, www.syracuse.com.

4 For a detailed discussion of the technical aspects of these processes, see Kruger, Inc., "The Power of Bundling: Syracuse, N.Y., WWTP Combines Technology to Address Discharge Concerns," *Wastewater Digest*, August 9, 2018, www.wwdmag. com.

5 Onondaga County Department of Water Environment Protection, "Metropolitan Syracuse Waste Water Treatment Plant Guide," www.ongov.net. Accessed July 15, 2020.

6 "SUNY ESF Professor Works to Restock Onondaga Lake with State's Once Abundant Atlantic Salmon," *Daily Orange*, June 22, 2010, http://dailyorange.com.

7 Honeywell: Onondaga Lake Cleanup, "Onondaga Lake Dredging and Capping Completed," www.lakecleanup.com. Accessed February 1, 2023.

8 "Roth Steel settles with DEC," Syracuse.com, April 18, 2008, www.syracuse.com. Accessed June 25, 2020.

9 "Lt. Gov. Hochul, DEC Commissioner Break Ground on Onondaga Lake Boat Launch," Auburnpub.com, April 25, 2019, https://auburnpub.com.

10 "Onondaga Lake Is Not Ready for Beach, Environmentalists Say," *Daily Orange*, October 27, 2019, https://auburnpub.com.

11 Ellen Abbot, "Proposed Beach along Onondaga Lake Would Cost $2.8M, Opponents Say the Lake Still Isn't Clean," March 2, 2020, www.wrvo.org.

12 "NYS DEC and U.S. Fish and Wildlife Service Announce New Opportunities for Public Input on Proposals to Restore Wildlife Habitat and Recreation on Onondaga Lake," State News Service, June 2, 2017, https://www.dec.ny.gov.

13 Department of Justice, "Honeywell to Restore Onondaga Lake Natural Resources under Proposed Agreement with the United States and the State of New York," Justice News, January 2, 2018, www.justice.gov. Accessed December 30, 2021.

14 Eric Higgs, *Nature by Design: People, Natural Process, and Ecological Restoration* (Cambridge, MA: The MIT Press, 2003), 14–58.

15 Stuart K. Allison, *Ecological Restoration and Environmental Change: Renewing Damaged Ecosystems* (London: Routledge, 2012), 6.

16 Higgs, 95.

17 Carolyn Merchant, *Reinventing Eden: The Fate of Nature in Western Culture* (New York: Routledge, 2003).

18 Rachel Carson, *Silent Spring* (New York: Houghton Mifflin, 2002).

19 J. Baird Callicott, *In Defense of the Land Ethic: Essays in Environmental Philosophy* (Albany: SUNY Press, 1989).

20 Robert Elliot, "Faking Nature: The Ethics of Environmental Restoration," *Inquiry: An Interdisciplinary Journal of Philosophy*, 25, no. 1 (1982): 81–93.

21 Allison, 10.

22 Allison, 22.

23 Allison, 27.

24 Allison, 36.

25 Allison, 37.

26 Higgs, 146.

27 Allison, 63.

28 Elizabeth Lev, Peter H. Kahn, Hanzi Chen, Garrett Esperum, "Relatively Wild Urban Parks Can Promote Human Resilience and Flourishing: A Case Study of Discovery Park, Seattle, Washington," *Frontiers in Sustainable Cities* (2020), www.frontiersin.org.

29 Allison, 73.

30 Ibid.

31 Higgs, 211.

32 Specific Claims Tribunal, Canada, www.sct-trp.ca. Accessed December 30, 2021.

33 "Let's Talk Indigenous Knowledge: Indigenous Knowledge Policy Framework for Proposed Project Reviews and Regulatory Decisions," The Government of Canada, www.canada.ca. Accessed December 28, 2021.

34 "Six Things You Need to Know About Bill C-69, Canada's Impact Assessment Act (IAA)," Golder: Member of WSP, August 26, 2019, www.golder.com.

35 Braela Kwan, "Indigenous Activists Spearhead Last-Ditch Effort to Short-Circuit Decades-Long Greenhouse Gas Boost," Investigate West, March 10, 2021, www.invw.org.

36 Lauren E. Eckert et al., "Indigenous Knowledge and Federal Environmental Assessments in Canada: Applying Past Lessons to the 2019 Impact Assessment Act," *Facets* (2020, February 13), www.facetsjournal.com.

37 JeanneTyler Moodee Lockman, "Over 1,000 Acres of Land Returned to Onondaga Nation for Creation of Wildlife Sanctuary," *CNYCentral*, June 29, 2022, https://cnycentral.com/news; U.S. Department of Justice, "Honeywell to Restore Onondaga Lake Natural Resources under Proposed Agreement with the United States and the State of New York," Justice News, December 20, 2017, www.justice.gov; Science Friday, "1,000 Acres of Ancestral Land Returned to Onondaga Nation," www.sciencefriday.com.

38 Wendy Gonyea, interviewed by author, October 5, 2022.

INDEX

Page numbers in *italics* refer to figures.

ABOUT THE AUTHOR

THOMAS SHEVORY is Professor Emeritus of Politics at Ithaca College. He is the author of many books, including *Toxic Burn: The Grassroots Struggle against the WTI Incinerator, Notorious H.I.V.: The Media Spectacle of Nushawn Williams*, and *The Great Lakes at Ten Miles an Hour: One Cyclist's Journey along the Shores of the Inland Seas*.

www.ingramcontent.com/pod-product-compliance
Lightning Source LLC
Chambersburg PA
CBHW020533030426
42337CB00013B/834